Running the Successful Hi-Tech Project Office

For a complete listing of the *Artech House Technology Management Library*, turn to the back of this book.

Running the Successful Hi-Tech Project Office

Eduardo Miranda

Artech House
Boston • London
www.artechhouse.com

Library of Congress Cataloging-in-Publication Data
Miranda, Eduardo, 1955–.
    Running the successful hi-tech project office / Eduardo Miranda.
      p.   cm. — (Artech House technology management and professional development library)
    Includes bibliographical references and index.
    ISBN 1-58053-373-6 (alk. paper)
    1. Project management.   2. High technology.   3. New products—Management.   I. Title   II. Series.
  HD69.P75M57   2003
  658.4′04—dc21   2002044072

British Library Cataloguing in Publication Data
Miranda, Eduardo
  Running the successful hi-tech project office. — (Artech House professional development and technology management library)
  1. Project management   2. High technology industries—Management
  I. Title
  658.4′04

  ISBN 1-58053-373-6

Cover design by Igor Valdman

© 2003 ARTECH HOUSE, INC.
685 Canton Street
Norwood, MA 02062

International Standard Book Number: 1-58053-373-6
Library of Congress Catalog Card Number: 2002044072

10 9 8 7 6 5 4 3 2 1

To Mariana, my wife, for her understanding and encouragement
To Tomás, my son, for making me proud every day
To my parents, for their love and example
Without all of them this book would never have existed

Contents

Acknowledgments

I would like to thank Anders Hemre, my boss at Ericsson Research Canada for his confidence and support; Lars Rosqvist from Bombardier Transportation; and Mats Hultin from Saab with whom I wrote my thesis on the project office for the Project Management Executive Master Program at Linköping University. I would also like to thank Stephen Hinton from Ericsson for reviewing the chapters and for his contribution to the Project Office Maturity Model; Raul Martinez from RMyA for reviewing the chapters and for challenging me all the way; and Gaetano Lombardi from Ericsson for his comments and review of the preliminary work. Finally, I would like to thank Michael Alton from Knowledge Based Systems for providing me with AIO Win, the tool used to produce the project office diagrams.

CHAPTER

1

Contents

Introduction

As organizations move toward a project approach as their preferred way of developing new products and services, the difficulty of maintaining aligned business strategies, resource needs, functional budgets, and delivery dates increases exponentially. Common symptoms of this misalignment are perpetual "fire fighting," weekly reprioritizations, and missed deadlines.

Many of the projects deemed failures in the project-management literature are not such, but the consequence of poor organizational policies or perverse modi operandi. Projects launched under seemingly good conditions soon become disasters when the necessary resources are not made available on time or promised deliveries from other projects are delayed. Under these circumstances, there is little that the project manager or his staff can do. The solution to chronic scheduling and budgeting problems requires not a new method of planning but avoidance of the conditions for failure.

In his book *The Logic of Failure* [1], Dörner states, "Failure does not strike like a bolt from the blue; it develops gradually according to its own logic. As we watch individuals attempt to solve problems, we see that complicated situations seem to elicit habits of thought that set failure in motion from the beginning. From that point, the continuing complexity of the task and the growing apprehension of failure encourage methods of decision making that make failure even more likely and then inevitable."

Illustrating how situational complexity leads to failure, Figure 1.1 compares the results of a simulation involving the

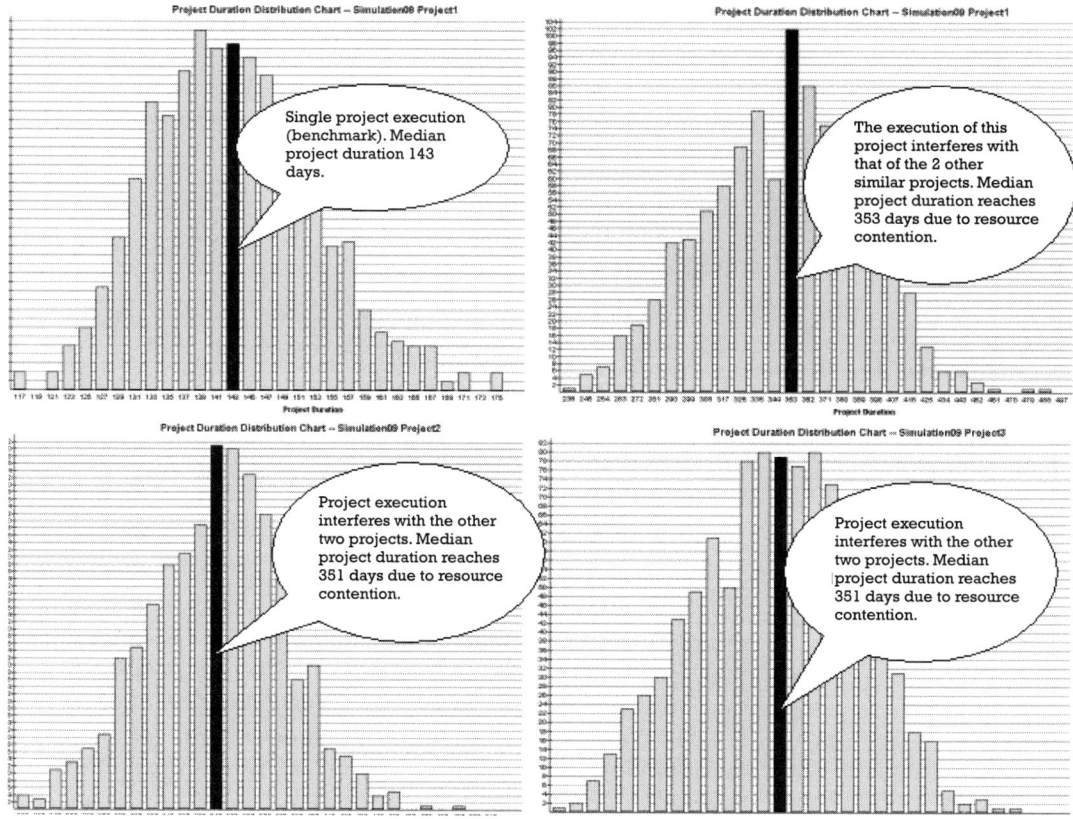

Figure 1.1 Project simulation results: *x*-axis shows project duration; *y*-axis shows number of occurrences for a given duration. (*After:* [2].)

simultaneous execution of three identical projects with the isolated execution of the same projects. The assumptions behind the simulation are few and realistic: Projects in the multiproject environment share resources, and prioritization is not guided by any particular policy but rather by whoever seems to be screaming the loudest at any given time. Also, there is a 1-day penalty each time any resource stops work in the middle of a project task in order to work on another project.

In the example shown, the project sponsors in the multiproject scenario must wait an average of 350 days to get their results instead of the 150 days corresponding to the promised duration of each project (see Figure 1.2). This should come as no surprise, since the work necessary for a simultaneous execution of the jobs exceeds the organization's capacity. However, by simply delaying the start of the projects by 20 days with respect to one

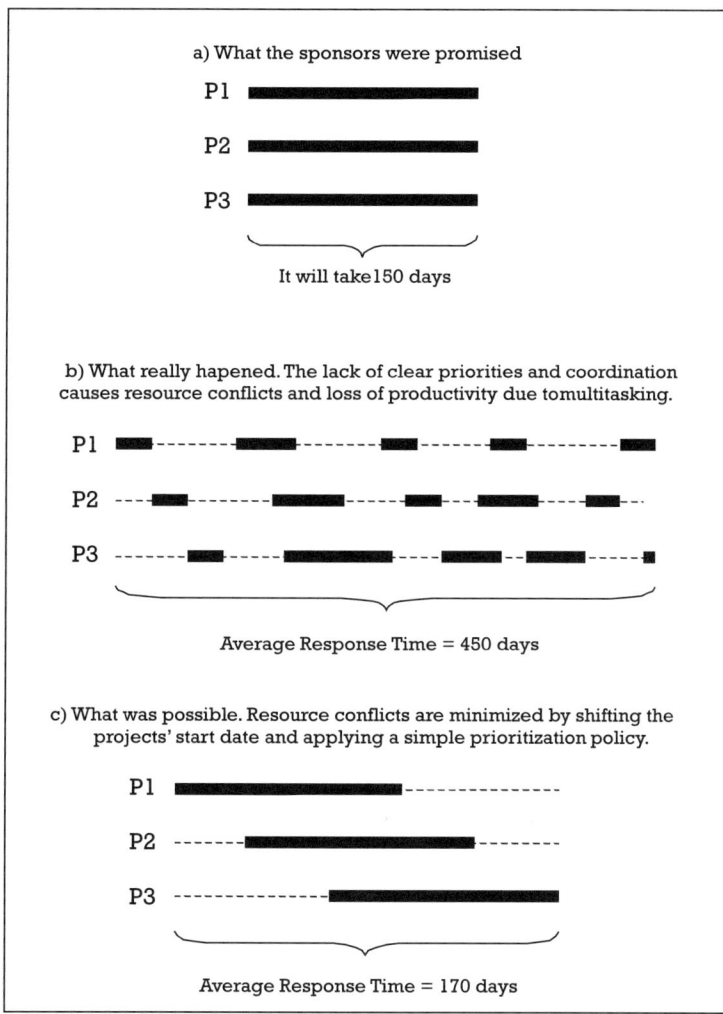

Figure 1.2 Expected results versus what really happened and what was possible.

another in order to minimize resource contention, and by giving priority to the projects according to when they were begun, the average waiting time could have been reduced to 170 days.

As this example shows, the need for coordination with respect to competing projects is self-evident. Although this fact seems both obvious and well understood, coordination problems plague the multiproject environment. I conducted an informal study within my own organization which found that 58% of the problems mentioned in risk and progress reports were due to lack of resources and insufficient coordination across projects (see Figure 1.3). Furthermore, in a study of 271 U.S. Department of Defense projects [3], respondents reported that technical problems accounted for only 25% of observed slips. Coordination problems such as funding stability, requirement changes, and "other causes" accounted for 75% of the delays.

1.1 Project portfolio management, project management, and line management

Project management, project portfolio management, and line management are three distinct functions; they have different purposes and time frames and require specific knowledge, authority, skills, and tools. Irrespective of the particular organizational form adopted—functional-, product-, project-, or matrix-oriented—all three functions are needed to accomplish an organization's goals. What changes from one organizational form to another is where the specific function is located and who has control over it. For example, in a matrix organization, the project management and line management functions will be performed by persons with different reporting lines, while in a functional organization they will be performed by the same or different people, but within a single scope of control.

	Internal Information LIST			2 (9)
Prepared (also subject responsible if other) XXXXXXXXXXXXXXXXXXXXXX	No. ZZZZZZZZZZZZZZZZZZZZZZZ			
Approved YYYYYYYYYYYYYYYYYYYYY	Checked	Date 2000-07-25	Rev F	Reference c

No	Risk Description	Corrective Actions	Risk Priority	Resp	Due Date	Status
1	No more pulling design resources for quick studies, rollout problems, from the project (sssssssss)	• Bring 15 CDC resources to LMC to be able to do market patches. • Tracking of resources pulled out for other projects/tasks. • Quick studies assigned to yearly resources. • Stronger maintenance organization. • Medium term: put into place a resource planning tool.	HIGH	xxxxxxx	6 April	Open Status: wwww to discuss with wwww. 13 April: ongoing 4 May: To compensate to elevate the delay we are taking resources from zzzzzzz project.
2	Need senior resources to do desk-check (five people).	• Re-plan design resources to allow for Code Review (CR). (Resource re-allocation) Function tester should be doing code reviews.	HIGH	yyyyyyyy	7 April	Status: Open 13 April: ongoing 4 May: Closed. Deskcheck has been completed by stopping the design work and review the use cases.
3						

Figure 1.3 Actual progress report showing typical coordination problems in multiproject environment (employee and product names have been changed).

Each function has different goals. For the project portfolio management function, this is to complete all projects to best achieve the business goals of the organization. For the project management function, the goal is to coordinate the work of a multidisciplinary team to produce a defined set of deliverables on-time and within budget. For the line management function, the goal is the timely provision of competent resources necessary to carry on the planned projects. Figure 1.4 shows the main relationships between them and with the rest of the organization.

The specific responsibilities laid out in this book for each of the functions are summarized in Table 1.1.

1.2 The project portfolio

An organization's project portfolio is the set of all projects, planned or in execution, formally acknowledged by the organization. Formal recognition usually comes in the form of the project clearing some type of gate put in place to separate the good projects from the not so good. At a minimum, the project scope will have been documented, the resource requirements estimated, the window of opportunity established, and the benefits to be derived from the project's execution evaluated so that an informed decision can be made regarding whether or not to go ahead with the proposal. Typical questions asked at the portfolio level include the following:

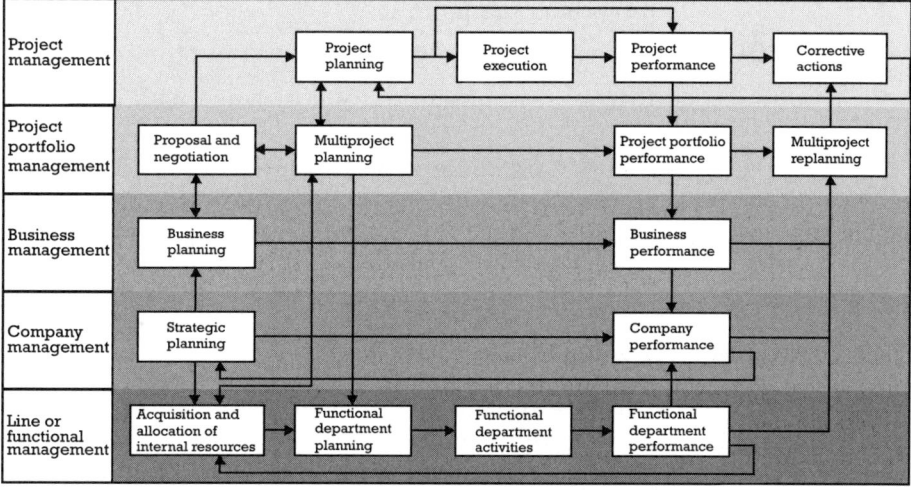

Figure 1.4 Relationships among project management, project portfolio management, and line management. (*After:* [4].)

Table 1.1   Proposed Organizational Responsibilities

	Project Management	Project Portfolio Management	Line Management
Time frame	Defined start and end	Ongoing	Ongoing
Focus	Manages team priorities to deliver within time, quality, and cost constraints	Manages workload to keep development pipeline operating efficiently	Develops organizational capability (resources, competencies, and processes)
Key decision areas	Prepares project schedule Prepares project budget Authorizes project expenditures Maintains communications with project sponsor Directs project team	Prepares long-term plans to enable appropriate decisions concerning the allocation of scarce resources Oversees the execution of projects on time and within budget Allocates project management personnel to projects Controls management reserves	Prepares line budgets. Staff development Determines staff compensation Recruiting and termination responsibility Owns functional process

- Do we have the resources necessary to execute project x in the time frame defined by its window of opportunity?
- What is the opportunity cost of project x?
- How does the cost-benefit ratio of project x compare with that of other projects in the portfolio?
- Can we rearrange the project mix in order to maximize the return?
- How can we offset the risk associated with project x?

The important thing about the project portfolio is that, similar to an investment portfolio, it allows the organization to select initiatives consistent with acceptable levels of return and risk.

1.3   The project office

The project office (PO) is an organizational entity chartered with supporting the work through projects. The extent of the PO support can range from serving as repositories of information regarding project-management best practices to being fully responsible for an organization's projects and programs.

The Gartner Group [5] defines three possible roles for the PO:

1. *The PO as repository:* At its most basic level the PO serves as the custodian for the project-management processes and as a repository for best practices. In this conception the PO is not formally involved in decisions concerning project execution; such decisions remain under the umbrella of the sponsoring functional areas.

2. *The PO as coach:* In its second organizational form, the PO takes a more active role. It provides guidance to project managers and participates in project reviews when called upon. The PO staff has a hand in the startup and closure of projects and provides expertise in specialized areas such as risk management and cost estimation when required. The PO performs some monitoring and consolidation of project performance reports, but does not order corrective action.

3. *The PO as manager:* In this role, the PO operates as an agent of senior management, which authorizes the PO to manage the project portfolio in its behalf. In addition to the responsibilities defined for the two previous roles, the PO is responsible for preparing a master plan and a resource plan reflecting the work the organization is committed to in the midterm, for reviewing project proposals, and for monitoring the execution of the portfolio.

Other organizations, such as the Dorsey Company [6], have produced similar classifications. Dorsey refers to them as *project support office, project/program office,* and *project governance office.* Table 1.2 summarizes the different types of POs.

Table 1.2   Project Office Types

Repository	Coach	Manager
Custodian of the organization's project-management processes (methods and tools)	Provides mentoring and coaching to project managers	Ensures alignment of projects and business strategy
Stores and disseminates best practices across projects and programs	Houses expertise in specialized areas such as estimation and risk management	Prepares the organization's master and resource plans
Provides a forum for communication	Provides analysis and reporting on project activity, such as status reporting and issues management	Reviews project scope and timing before including it in the portfolio
Supports project managers with guidance and expertise with respect to project-management practices	Consolidates reporting across the organization	Orders corrective actions for individual projects.
	Audits projects and programs at request of the project sponsor	Assigns project managers

The different types of POs must not be confused with maturity levels or stages of institutional development. An organization with a manager type of PO is not necessarily better or more mature than one deploying a repository type of PO. The different types reflect different organizational needs, cultures, and governance realities.

This book develops on the idea of the PO as manager. Those who need a less powerful type of implementation could use the responsibility matrix described in Chapter 3 (see Table 3.1) to tailor-down proposals to the needs of their organizations.

1.3.1   Making the case for the PO

Although the decision to deploy a PO in a given organization needs to be made on a case-by-case basis, it is possible to examine what is at stake by looking at the research and development (R&D) expenditures reported by several well-known companies. The reason for choosing R&D expenditures to exemplify the potential savings and gains that could be realized through a PO is threefold: First, the numbers are publicly available in the annual reports of the companies. Second, most R&D work adopts the form of

Table 1.3   R&D Expenditures

Company	Industry	R&D Expenses for 2000 (in millions of dollars)	Assuming Operations Improve by*		
			1%	5%	10%
Ericsson	Communication equipment	4,014	40	200	401
Nortel	Communication equipment	4,005	40	200	400
Microsoft	Software	3,772	37	188	377
Merck	Pharmaceutical	2,343	23	117	234
Novartis	Pharmaceutical	1,984	19	99	198
Oracle	Software	1,000	10	50	100
J.D. Edwards	Software	116	1.1	5.8	11.6
Rational Corp.	Software	102	1	5.1	10.2
Ballard Power	Fuel cells	55	.5	2.8	5.5
Cognos	Software	54	.5	2.7	5.4
Borland	Software	42	.4	2.1	4.2
Activision	Entertainment	26	.26	1.3	2.6
Eaglepoint	Software	3	0	.15	.3

*Results in savings in millions of dollars per year.

projects. Third, R&D organizations exhibit most of the problems we will be referring to throughout this book.

Table 1.3 shows the annual savings that could be realized by these companies, should the implementation of a PO bring a lasting improvement of 1%, 5%, and 10% of their operations. The savings likely to be brought by the implementation of a PO are the result of a better coordination of resources, consistent decision making across projects, and risk sharing among all of the projects in the portfolio.

For those who think that it is not possible to realize savings of these magnitudes simply by improving the project-management function, the charts in Figure 1.5 should suffice to show the influence that project management can exert over development costs. The first chart in the figure shows development costs attributed to managerial factors in a study of R&D work in the pharmaceutical industry; the second one shows the waste incurred in the information technology sector based on a study by the Standish Group, and the third the effort wasted by organizations according to their portfolio-management practices in a study conducted by the firm PRTM.

But POs are not just about cutting costs and creating shareholder value. The PO can also have an impact on the quality of life of employees working in R&D environments.

The 1990s was a decade of change. Companies were making money despite themselves. At the height of the dot-com fever, being first to market was the be-all and end-all, and performance targets were set with little or no connection to organizational capacity; as a consequence, work–life conflicts increased markedly [10]. Employees were, and still are, putting in longer hours at work and experiencing greater challenges in balancing their roles as employee, parent, spouse, and community member. As Figure 1.6 shows, workers have become more stressed and physical and mental health has declined, along with personal satisfaction.

It is undoubtedly essential that a corporation have a healthy bottom line—after all, its employees' jobs depend on it—but must a healthy bottom line come at the expense of employee quality of life? Furthermore, when an employee is putting in 60 hours a week, week after week, is he or she really making a valuable contribution? Is it sustainable? Are we really doing more, or only spending more time at work? The model depicted in Figure 1.7 and the statistics in Table 1.4 show the negative consequences of work and family stress.

As will be shown in later chapters, by bringing the organizational workload under control and eliminating the need for constant fire fighting, the PO can positively affect individual—and corporate—well-being.

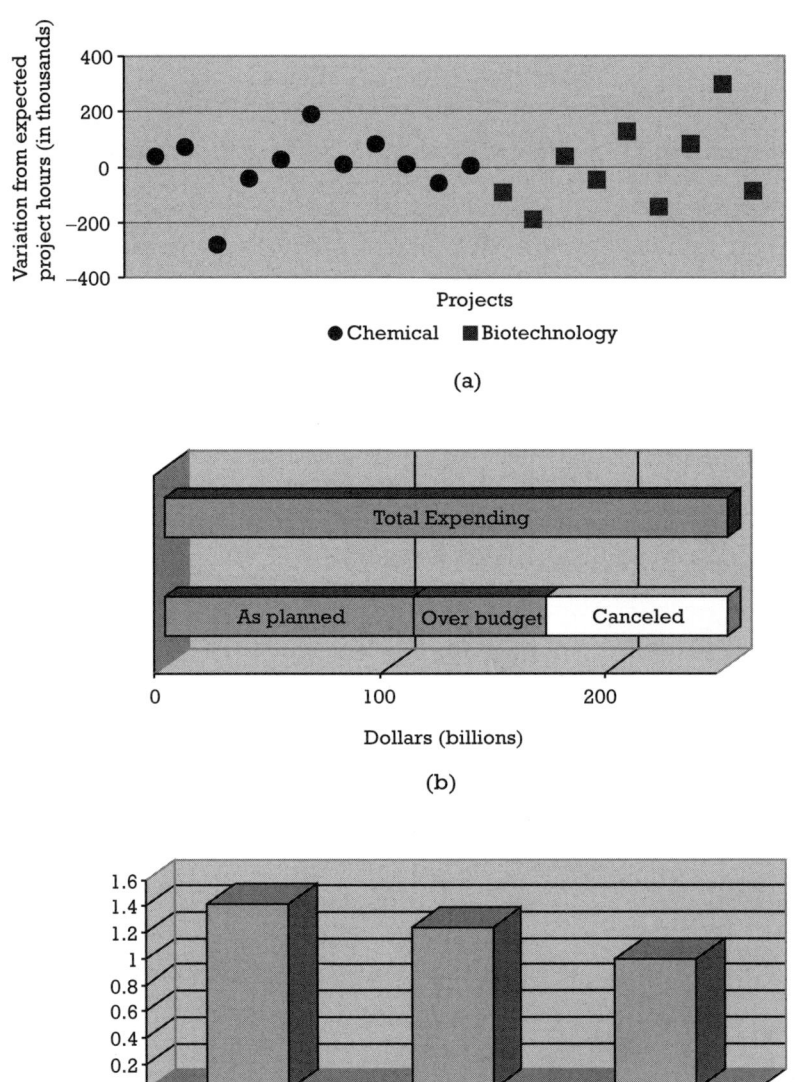

Figure 1.5 Effects of project management on development costs. (a) Management influence on cost: The y-axis shows the difference between the actual and the expected hours spent in the development of new pharmaceutical products after discounting project complexity, production scale, and therapeutic class. (b) Waste on IT projects: Almost one-third of the total spending, billions, never comes to fruition. (c) Organizations where projects are executed in functional silos waste, on average, 42% more effort than organizations that use the portfolios approach owing to inconsistent decision making concerning the termination of failed projects. (*After:* [7–9])

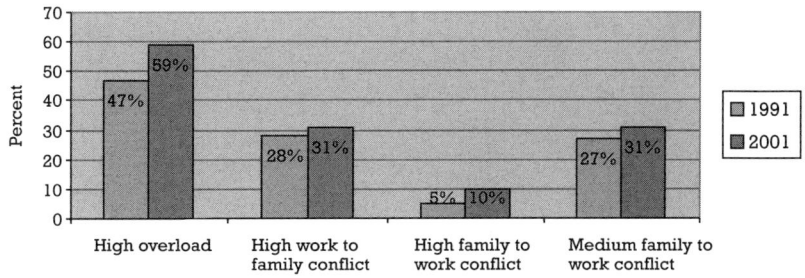

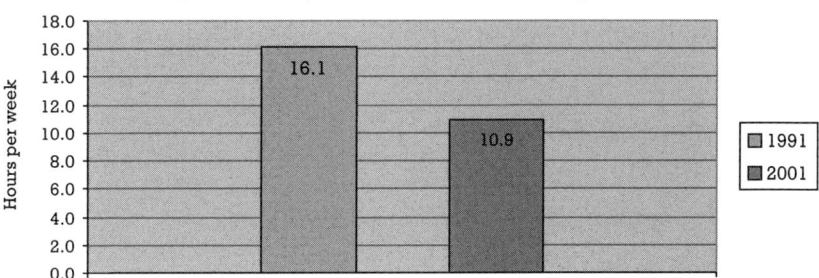

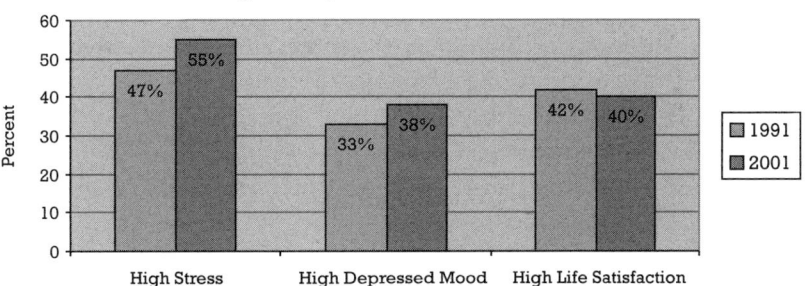

Figure 1.6   Quality-of-life indicators, 1991 versus 2001. (*After:* [10].)

1.3.2   Obstacles to successful deployment

To be successful, the deployment of a PO must be done in accordance with the culture, requirements, and governance realities of the organization. This is unmistakably indicated by the disparity of the responses given in a survey [11] inquiring about the main obstacles facing the deployment of a PO in three Swedish companies. Although most of the respondents agreed that

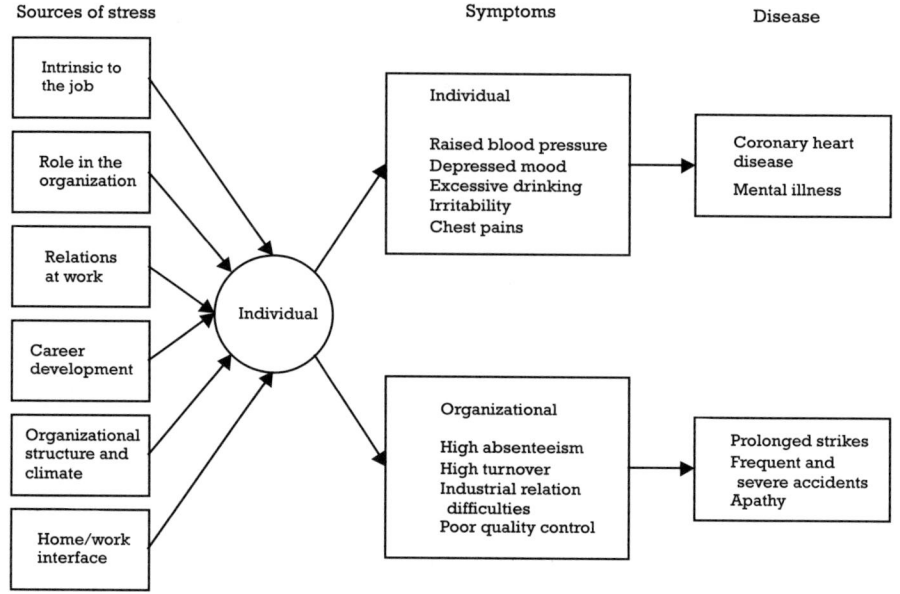

Figure 1.7 Dynamics of work stress. (*After:* [11].)

having a managerial PO could be a solution to their multiproject problems (see Figure 1.8), they perceived very different obstacles as potential hindrances to such an entity's successful deployment (see Figure 1.9). This disparity is probably a reflection of the experience, career interests, and position of the respondents within their respective organizations.

A successful implementation must thus begin with an assessment of an organization's current situation and the adaptation of any generic model, like the one proposed in this book, to the realities of the organization.

1.4   Summary

The last decade has seen an impressive growth in the number of organizations using the project approach to conduct their business. But this growth, explained by the effectiveness and flexibility of the project work form in an environment of increasing complexity and demands for "faster, better, and cheaper," has created problems of its own: conflicts between projects competing for the same resources, lack of coordination between complementary initiatives, and loss of sight of the organization as a whole. Coincidentally, there has been an increase in the number of work hours, the level of stress, and the number of work–family conflicts experienced by employees.

Table 1.4 Consequences of high work/life conflict (*After:* [10])

Consequence	Type and degreet of conflict					
	Role Overload[1]		Work To Family[2]		Family To Work[3]	
	High	Low	High	Low	High	Low
Lack of organizational commitment (%)	53	42	58	39		
Lack of job satisfaction (%)	70	40	80	40	67	46
High job stress (%)	44	2	57	9	20	40
High employee turnover. Think of leaving weekly. (%)	18	7	26	7	20	7
Missed work time due to family (%)	54	37	55	45	75	42
Absenteeism due to:						
Health problem (%)	54	41	54	47	60	46
Family problem (%)	29	12	28	20	55	10
Mental health day (%)	39	18	40	26	40	28
Average number of days absent in six months due to:						
Health problem (days)	3.7	2.7	3.5	3.0	4.0	3.0
Childcare problem (days)	0.7	0.3	0.6	0.4	1.6	0.3
Eldercare problem (days)	0.3	0.2	0.7	0.3	0.5	0.3
Mental health day (days)	0.9	0.2	0.9	0.2	1.0	0.2
Total (days)	5.6	3.4	5.7	3.9	7.1	4.2
Rate organization as above average place to work (%)	48	75	39	72	48	60
Rate organization as below average place to work (%)	13	4	19	4	12	9
Use EAP/psychological counseling (%)	75	5	35	24	35	8
Purchased prescription medicine (%)	65	54				

1.Role overload exist when the total demand on the time and energy associated with a person's responsibilities in different roles, such as parent, community member, employee, are too great to perform the roles adequately or comfortably.
2.Work-to-family conflict occurs when work demands make it more difficult to fulfill family responsibilities.
3.Family-to-work conflict occurs when family responsibilities make it more difficult to fulfill work responsibilities.

     The same set of circumstances reported across organizations and industries points to a systemic or structural, rather than a performance problem. This book proposes the creation of a new business function, the PO, as the means of coordinating, managing, and reporting on projects across the organization.

     Introducing a PO into an organization is a substantial undertaking, and it is essential to remember that all change, especially change involving the entire organization, takes time. The establishment of the infrastructure necessary to support the PO is only part of the solution. The organization must allow its culture to evolve—and it must be prepared to do business in a different way—in order to succeed.

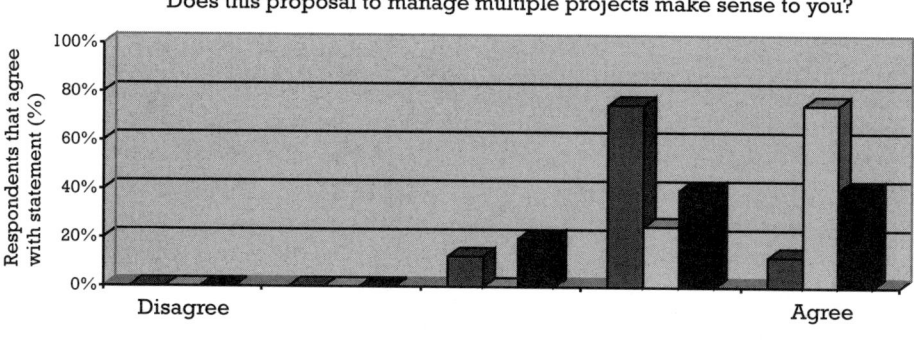

Figure 1.8 Survey results showing positive perception of PO concept. (*Source:* [12].)

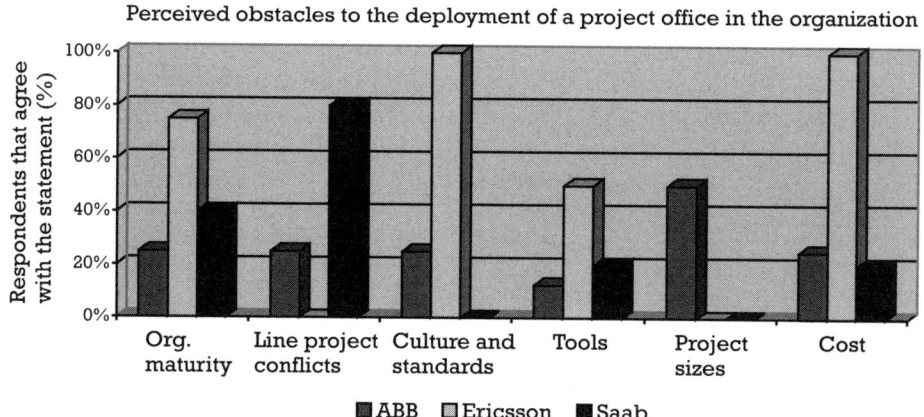

Figure 1.9 Perceived obstacles to PO deployment. (*Source:* [12].)

References

[1] Dörner, D., *The Logic of Failure: Recognizing and Avoiding Error in Complex Situations*, Reading, MA: Addison-Wesley, 1997.

[2] Goldratt, E., *Project Management the TOC Way*, Croton-on-Hudson, NY: North River Press, 1998.

[3] McNutt, R., "Reducing DoD Product Development Time: The Role of the Schedule Development Process," Ph.D. diss., Cambridge, MA: Massachusetts Institute of Technology, 1998.

[4]   Cagno, E., et al., "Project Prioritization in a Multi-Project Environment," *IPMA 14th World Conf. on Project Management*, Ljubljana, Slovenia, 1998.

[5]   Light, M., *The Project Office: Teams, Processes, and Tools*, Gartner Research, 2000, http://www.techrepublic.com.

[6]   Belzer, K., *The Program Office: A Business Results Enabler*, R. Dorsey & Company, http://www.pmforum.org/library/papers/ProgramOfficeFinal.htm.

[7]   Pisano, G., *The Development Factory: Unlocking the Potential of Process Innovation*, Boston: Harvard Business School Press, 1997.

[8]   The Standish Group, "The Chaos Report," 1994, http://www.pm2go.com/sample_research.

[9]   Pittiglio et al., *Recipes for Growth in Technology-Based Industries*, 1998.

[10] Duxbury, L., and C. Higgins, *Work–Life Balance in the New Millennium: Where Are We? Where Do We Need to Go?* CPRN Discussion Paper No. W/12, October 2001, http://www.cprn.org.

[11] Cooper, C. L., and J. Marshall, "Occupational Sources of Stress: A Review of the Literature Relating to Coronary Heart Disease and Mental Health," *Journal of Occupational Psychology*, Vol. 49, 1976, pp. 11–28.

[12] Miranda, E., L. Rosqvist, and M. Hultin, *Managing Multiple Projects: The Project Office Organization, Roles, Responsibilities, Processes and Tools*, Linköping, Sweden, University of Linköping, 2001.

2

Contents

The multiproject challenge

2.1 Introduction

In a multiproject environment, decisions made within one project tend to affect other, seemingly unrelated, projects—even those not yet in execution.

The model shown in Figure 2.1 illustrates how the consequences of local project decisions, such as raising the level of overtime or delaying the beginning of a task, ripple through the project portfolio via invisible links created by the use of shared resources. The model is based on my own observations over many years—first as a software developer and then as a project manager—as well as the observations of my colleagues and the extensive project modeling work done by Cooper [1], Abdel-Hamid and Madnick [2], and Sterman [3]. The basic assumptions behind the model are as follows:

▸ Even the best-planned projects are plagued with uncertainty.

▸ Projects in a portfolio interfere with one another.

▸ The multiproject environment is a complex behaved system.

2.1.1 Project uncertainty

Uncertainty means variability, and variability is the fundamental rationale behind project management. If there were no variability in the time or effort required to complete a given task, the project approach would be simple: One would simply devise good plans and then execute them. But no matter how good the plan, the estimates on which schedules and resource

17

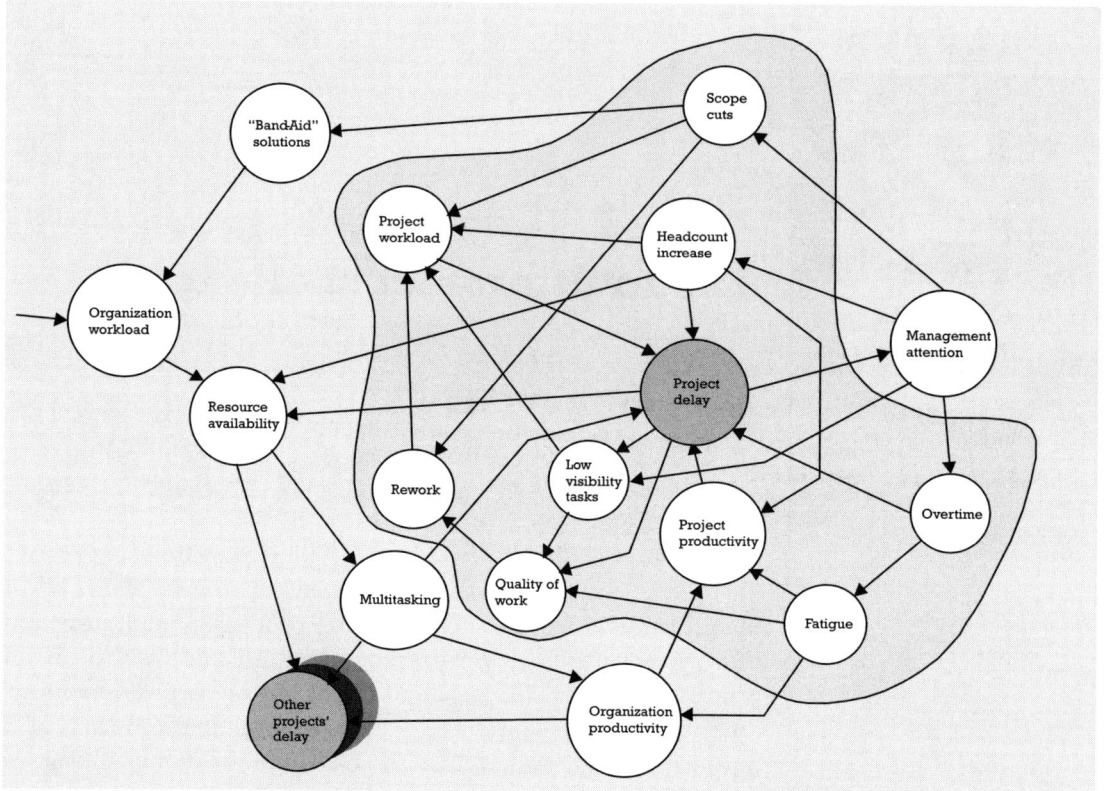

Figure 2.1 Relationships among principal management variables in a multiproject environment.

allocations are based rest on a multitude of assumptions with respect to task complexity, worker ability, the ability of suppliers to deliver on time, the availability of resources associated with other projects, and even unknown unknowns (see Figure 2.2).

A project will finish on time or late depending on which of these assumptions turns out to be valid or invalid during project execution and on how those involved choose to react. The use of probability distributions, such as the beta distribution and the triangular distribution in critical-path calculations, is intended to capture these facts.

2.1.2 Project interference

In the modern project-based organization, projects have explicit and implicit links. These links originate in the sharing of resources and results among

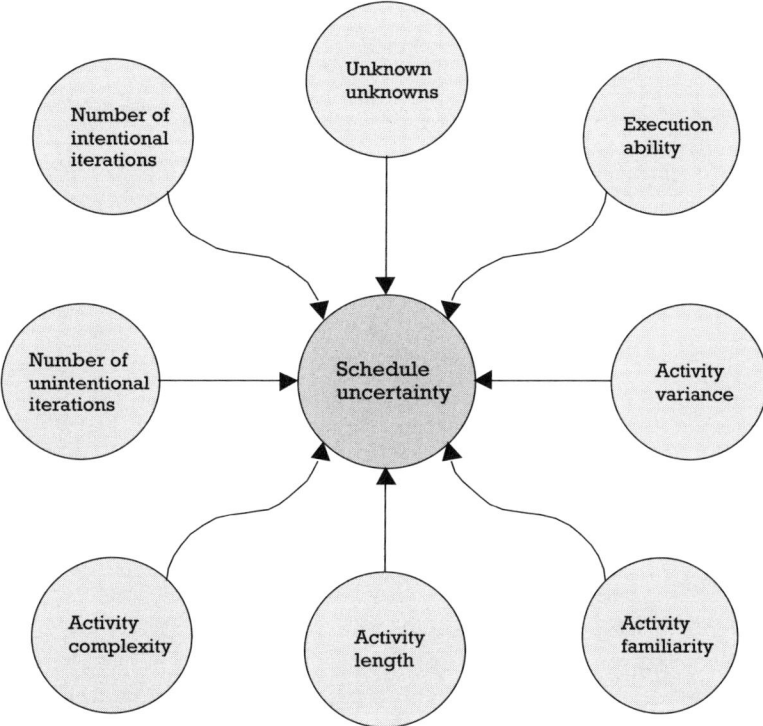

Figure 2.2 Sources of scheduling uncertainty. (*After:* [4].)

projects, extending to other projects in the portfolio the consequences of actions taken within one project. It is for this reason that decisions that seem to make sense in the context of one project might not be such a good idea when considered in light of the entire portfolio. As shown in Figure 2.3, the number of potential interactions among projects increases geometrically with the size of the portfolio and the number of line functions involved in project execution.

2.1.3 Complex behaved system

A complex behaved system is a system whose responses are nonlinear, time-lagged, and time-dependent. The whole is not quantitatively equal to its parts, or even qualitatively recognizable in its constituent components. Results cannot be assumed to be repeatable; the same experiment may not come out exactly the same way twice, as outcomes are not only a function of the input conditions but also of the time at which they occur.

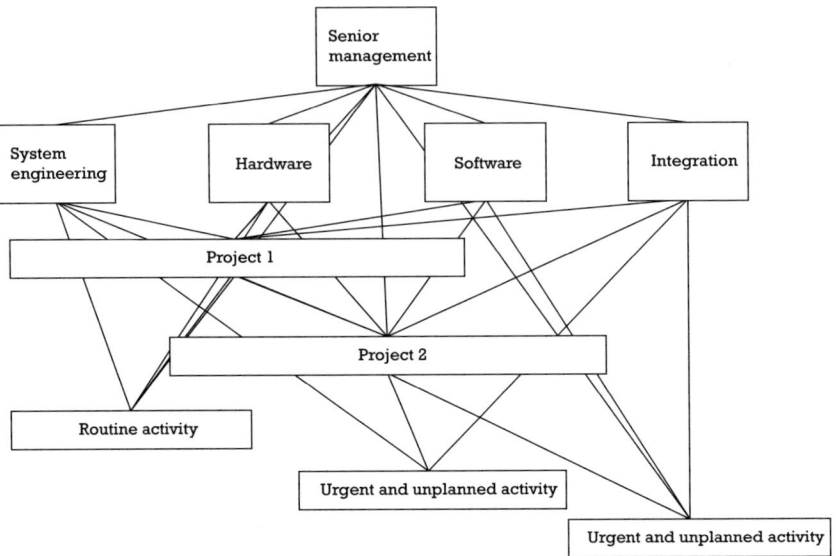

Number of potential interactions is $N(N\text{-}1)/2$, $N = 10$ (functions + projects + other activities), potential interactions = 45

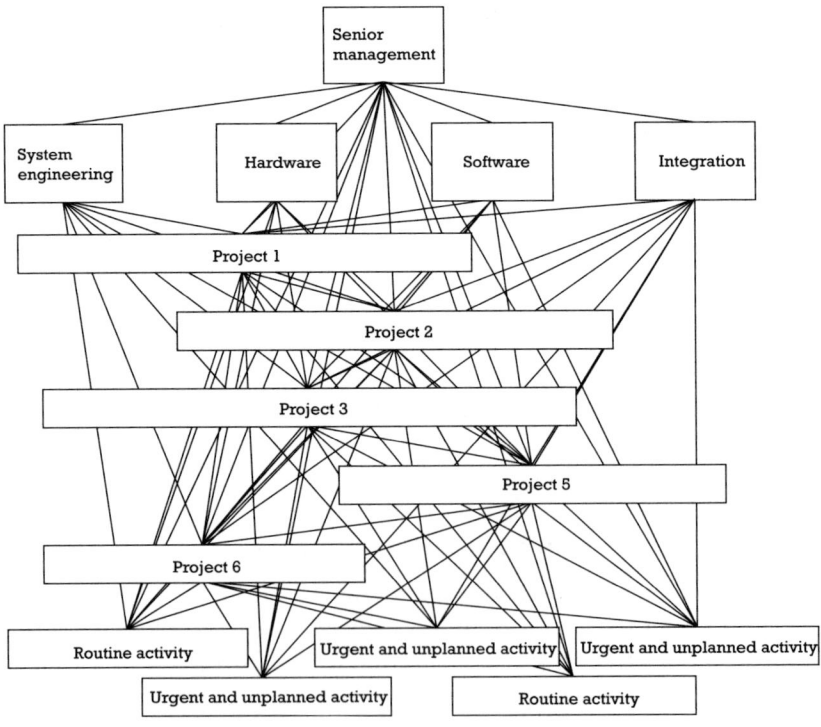

$N = 15$ (functions + projects + other activities), potential interactions = 105

Figure 2.3 Interactions in the multiproject environment.

Complex behavior can be seen in any system made up of a large number of interacting constituents, be they atoms in a solid, cells in a living organism, consumers in a national economy, or in our case, projects, resources and sponsoring organizations. But in the multiproject environment, the sheer number of interactions is not the only source of complex behavior; the feedback loops that exist among the principal management variables, and the fact that time is an independent variable, also play a role.

The problem with systems exhibiting complex behavior is that they cannot be steered in the desired direction by applying any single action at any given time—in the case of a multiproject environment, increasing the head count in one project means delaying the start of another, increasing the use of overtime means diminishing the productivity of the whole organization, and not fixing some defects now means fixing them later at a higher cost. There is also an omnipresent risk of "oversteering" the system.

2.2   The multiproject environment

The diagram in Figure 2.1 depicts the principal variables upon which management tends to act in an attempt to keep projects on track. Such variables tend to address the current situation, as long-term approaches such as process improvement and competence development are seldom a remedy in the case of late projects.

The arrows in the diagram describe influence relationships among the management variables. For example, overtime hours lead to worker fatigue, and fatigue affects productivity, leading to project delay. When a project is delayed the project team shifts its focus from low-visibility tasks, such as inspections, reviews, and testing, to high-visibility tasks, such as coding and integration; this causes an immediate reduction in the project workload and some of the delay is recouped. Unfortunately, errors that would otherwise have been caught through these low-visibility tasks have moved to the next stage in the project's development, at which time the cost to fix them is of an order of magnitude greater than the time originally recouped by eliminating them. As the schedule pressure continues to mount, the quality of the decisions made by the project team deteriorates, the number of errors increases, and the project falls even further behind. At some point, the project begins to attract special management attention, and the project staff is asked to work harder on "value-adding" activities.

The team gets the message, and begins to put in longer hours while the focus on quality-oriented activities continues to drift away. The extra hours soon pay off in the form of boosted output, but as people become fatigued, their productivity drops and the number of mistakes made increases, creating another vicious circle.

The next weapon in the conventional management arsenal is to increase the project head count. This measure could help or damage the project, depending on the circumstances. We know for a fact that some effort is associated with taking the newcomers through the learning curve, and this implies an increase in project workload for the original, already overloaded, staff. To add to this, through the learning period, the new staff is far from being 100% productive, which lowers the average team productivity. Furthermore, if the work was not planned from the beginning to accommodate the extra head count, a significant effort might be needed to partition and later integrate the new interfaces. Also, as the team grows larger, its productivity diminishes as a result of an increase in the communications overhead and process losses [5].

Finally, after the above approaches fail to produce the desired result, the project scope is reduced. This, which effectively cuts the workload, also comes with a price tag: Interfaces must be reworked and adaptations made. In the end, this may result in less product functionality and no real savings.

Outside the sphere of the offending project, other otherwise unrelated projects begin to experience delays, as the resources they need are not made available on time. Projects waiting for deliverables are also affected. As the number of projects in the queue increases, resources are multitasked across several projects in order to keep the project sponsors happy. Such ad hoc multitasking further hinders the overall productivity of the organization. To add to the mayhem, limiting the scope of the offending project has resulted in a number of Band-Aid projects intended to pacify customers who were promised now defunct features. This adds to the organizational workload, further reducing the resource availability and increasing the multitasking, which further reduces the productivity, which further delays the projects, which adds to the workload, which further reduces the productivity. Breaking the circle calls for something truly dramatic.

2.3 Self-fulfilling prophecies

The multiproject system described above displays the characteristics of a complex system in the following respects:

> ▸ *High coupling:* Any intervention is likely to affect something somewhere else.

▸ *Time lagging and nonlinear responses:* The results of an intervention take time to materialize and when they do, they do not materialize at a constant rate of progress.

▸ *The true state of the system is unknowable:* The current state of the system can only be inferred, its most likely evolution only guessed at. The presence of many actors, each with his or her own agenda and opinion about what should be done and when, and whose behavior is conditioned by the behavior of others, contributes to the fact that the system resists reductionistic analyses.

Thus, the multiproject system is inherently unstable; the best advice one can give on how to manage it is to avoid getting into trouble in the first place, because once one gets trapped in the vicious circle it is very difficult to get out.

Repenning, Gonçalves, and Black [6] argue that every organization has a tipping point, a threshold that determines how much development and how much problem fixing an organization can handle, which once crossed causes fire fighting to spread rapidly from a few isolated projects to the entire organization. The cornerstone of their theory is that the more up-front work done in a project, the less difficulties encountered downstream (see Figure 2.4). Based on this model, Repenning, Gonçalves, and Black produced the chart shown in Figure 2.5. In this figure we see that the development system behaves either like a virtuous or a vicious circle, depending on how much up-front work is done in the current year in support of next year's projects.

So other things being equal, in a fire-fighting situation desperate interventions lead to more desperate measures, which justify bringing in more firefighters, which leads to more fire fighting, which leads to more extreme measures.

2.4   Common responses to project delays

The following sections provide the factual basis for the claims made in the preceding section. Much of the data used has not been drawn from the project environment—such data is either nonexistent or difficult to access. However, to jump from the original fields of research where the data was collected to project work is not too much of a stretch. For example, if it has been established that air-traffic controllers make more mistakes when they are tired, doesn't that corroborate the contention that people working in projects under high-stress conditions make more mistakes? If psychological

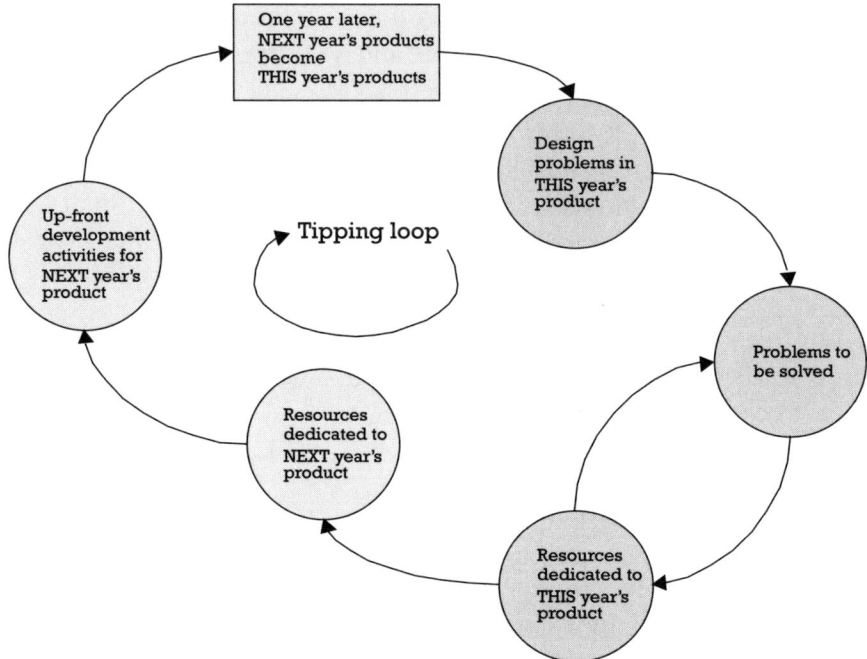

Figure 2.4   Feedback structure of multiproject product development. (*After:* [6].)

studies show that students forget half of what was said in class after only a day or two, can't we make assumptions about what happens to an engineer's train of thought when she is interrupted to work on a more pressing task?

2.4.1   Cutting back or eliminating low-visibility tasks

When a project begins to fall behind schedule, one of the first things to suffer are low-visibility activities, such as tradeoff studies, inspections, and reviews, which are curtailed or dropped altogether in favor of high-output activities, such as coding and integration. In general, people in situations in which they must achieve multiple goals, such as meeting a deadline and achieving a certain level of quality, tend to sacrifice the least visible goal when they perceive that satisfying both would be difficult. In a study conducted by Gilliland and Landis [7] which attempted to evaluate the tradeoffs between quality and quantity when both goals were perceived as being difficult to attain simultaneously, participants gave up the less visible quality goal (see Figure 2.6) and allocated their efforts toward the more achievable

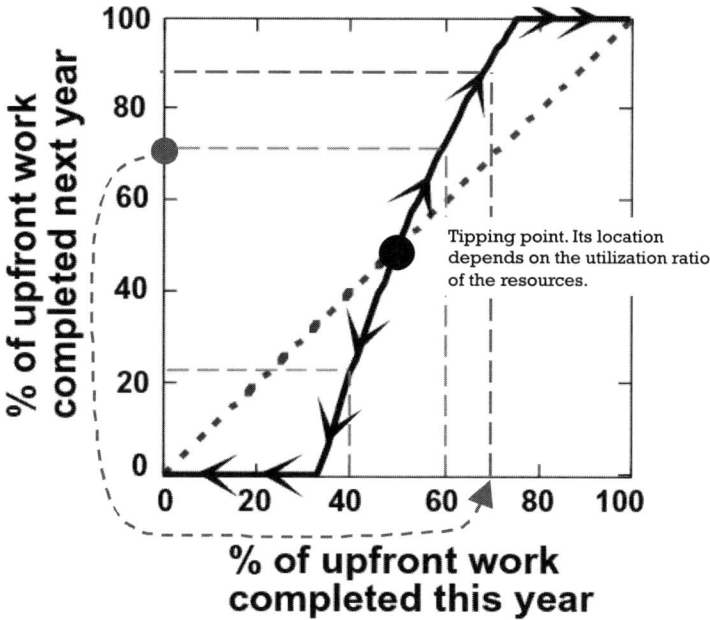

The tipping point represents the equilibrium point. The arrows show the direction in which the system will evolve when in disequilibrium. The dashed lines show how to read the diagram. In the example the 60% of upfront work this year will enable the organization to do 70% of upfront work next year and 80% the following and so on. This is the virtuous cycle. If the upfront work done this year is 40%, next year the organization will be able to do only about 23%, the rest of the time will be spent fixing problems originated in the previuos year. This is the vicious cycle. If the trend continues, soon all the organization is operating in firefighting mode.

Figure 2.5   Execution modes in a multiproject environment. (*After:* [6].)

quantity goal. Weinberg and Schulman [8] came to a similar conclusion (see Table 2.1). In their experiments, five teams were given the same programming assignment but different goals to achieve. The experiment showed two remarkable results:

1.   Four of the five teams excelled with respect to the objective they were given. One finished second.

2.   None of the teams performed consistently well with respect to all of the objectives.

So if the organization, explicitly or implicitly, favors one goal over another, when something has to give, rest assured it will be the less favored, less visible goal.

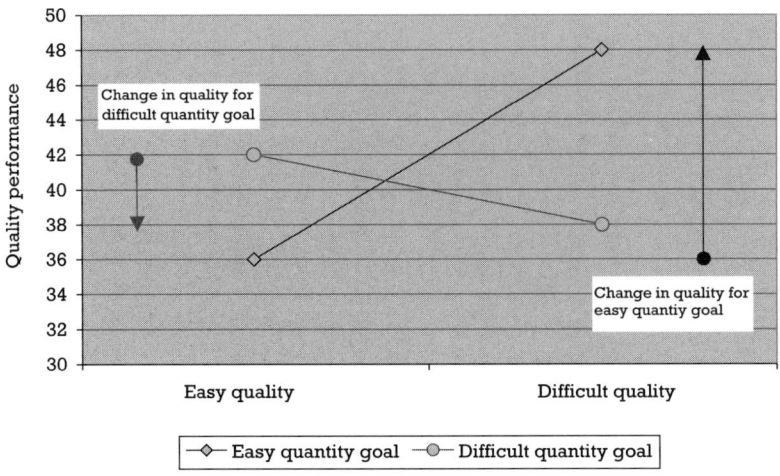

The Gilliland and Landis experiment shows that when the quantity goal wa easy to achieve, setting a difficult quality goal resulted in increased quality, but when a difficult quantity goal was set (the participants had little time to think about quality) assigning a difficult quality goal not result in an improved quality score.

Figure 2.6   Quantity over quality. (*After:* [7].)

Table 2.1   Results of Weinberg-Schulman Experiment

Team's Objective	Ranking with Respect to All Objectives				
	Effort To Complete	Number of Statements	Memory Required	Program Clarity	Output Clarity
Effort to complete	1	4	4	5	3
Number of statements	2–3	1	2	3	5
Memory required	5	2	1	4	4
Program clarity	4	3	3	2	2
Output clarity	2–3	5	5	1	1

In a larger setting, the Software Engineering Institute (SEI), in its 2001 report on process maturity [9], stated, "Software Quality Assurance is the least frequently satisfied Level 2 Key Process Area among organizations assessed at Level 1" (see Figure 2.7). In Figure 2.7, we can see that peer review, an effective but low-profile activity, is seldom practiced even among those organizations looking to be assessed at Level 3 of the SEI maturity scale. In its study of NASA's project management practices (see Table 2.2), the Mars Climate Orbiter [10] Mishap Investigation Board traced many of

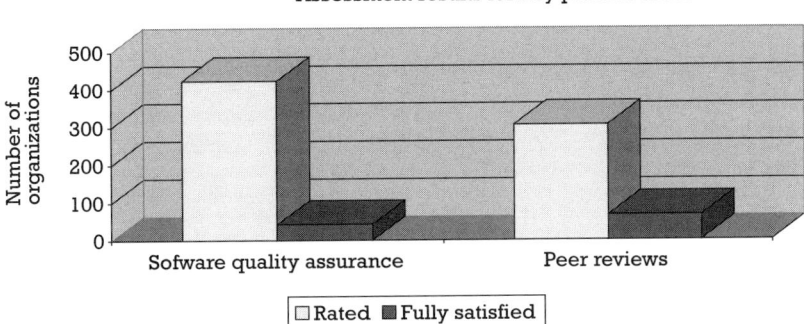

Figure 2.7 Software quality assurance and peer reviews are among the least satisfied key process areas of the Software Engineering Institute's Capability Maturity Model. (*After:* [9].)

the mission's problems to the nonperformance of reviews and risk-management activities.

2.4.2 Effects on product quality and decision making

Many factors affect the quality of work, but here we are interested in the drop in quality resulting from the abandonment of reviews, inspections, regression testing, and other error-identification activities, and from the deterioration in the quality of the decisions made as a result of increased time pressure.

We saw in the previous section that low-visibility activities such as inspections and risk management tend to suffer when they conflict with other tasks of a more visible nature. To understand the negative impact this has in the project workload, it is first necessary to understand the value of inspections, code reviews, and regression testing. Table 2.3 shows the effectiveness of various quality activities.

When these activities are abandoned, the errors that they could have detected will move to later stages in the product life cycle, where they will have to be fixed at a higher cost. Of course, the most expensive errors are those uncovered only after the product has been released for general use—or in the case of NASA, after the spacecraft has been launched.

Similarly, time pressure affects the quality of the decisions made. In a study conducted by Kim, Wilemon, and Schultz [12] about the consequences of stress on new-product development projects, time pressure was found to be one of the main stressors, and its effect on the quality of decisions and interpersonal relationships mostly negative. When asked to what

Table 2.2   Recurring Themes from Failure Investigations and Studies at NASA

Project Theme	Mars Climate Orbiter	Wide-field Infrared Explorer	Lewis	Boeing MAR	Faster, Better, Cheaper	Solar Hello-spheric Observatory	LMA IAT on Mission Success	Space Shuttle IA Team	Frequency
Reviews	X	X	X	X	X			X	6
Risk management	X		X	X	X	X		X	6
Testing, verification, validation	X	X		X		X	X	X	6
Communications	X		X			X	X	X	5
Health monitoring during critical ops	X	X			X				3
Safety and quality culture	X			X			X		3
Staffing	X				X				2
Continuity	X				X				2
Cost and schedule			X		X				2
Engineering discipline			X	X					2
Government/contractor roles and responsibilities			X					X	2
Human error							X	X	2
Leadership	X					X			2
Mission assurance	X				X				2
Overconfidence	X							X	2
Problem reporting	X							X	2
Subcontractor, supplier oversight				X			X		2
System engineering	X		X						2
Training	X						X		2
Configuration control						X			1
Documentation					X				1
Line organization involvement	X								1
Operations	X								1
Project team					X				1
Requirements			X						1
Science involvement	X								1
Technology readiness					X				1
Workforce stress								X	1

After: [10].

extent stress affected their work performance, 69% of the participants responded that it adversely affected them to some extent or to a great extent (see Table 2.4).

Table 2.3   Contribution of "Low Visibility" Quality Activities in the Software Industry

Quality Activity	Defect-Identification Effectiveness (%)
Informal design reviews	25–40
Formal design inspections	45–65
Informal code reviews	20–35
Formal code inspections	45–70
Unit test	15–50
New function test	20–35
Regression test	15–30
Integration test	25–40
System test	25–55
Low-volume beta test (10 clients)	25–40
High-volume beta test (1,000 clients)	60–85

After: [11].

Table 2.4   Impact of Stress on Performance

Impact of stress on performance	Number of respondents	Percentage
Very great extent	5	8.6
Great extent	11	19.0
Some extent	24	41.4
Small extent	16	27.6
Not at all	2	3.4

After: [12].

Table 2.5   Common Barriers to Thinking

Ranking by Workers	Ranking by Managers
Organizational politics	Time pressure
Time pressure	Organizational politics
Lack of involvement in decision making	Lack of involvement in decision making

After: [13].

In a survey of 1,414 employees (641 managers and 773 hourly workers) conducted by the consulting firm Kepner-Tregoe [13], the most common barriers to thinking reported by the respondents (see Table 2.5) were organizational politics, time pressure, and lack of involvement in decision making.

Another study conducted by the same firm [14] shows that one of the ways in which employees deal with time pressure is by making less quality decisions (see Figure 2.8).

2.4.3 Rework

Many studies have been conducted that compare the cost of fixing or avoiding a problem early on versus fixing it at a latter time. Although the results are somewhat disparate, they are consistent in finding that the cost of a late fix is higher than that of an earlier fix. Figure 2.9 summarizes the results of some well-known studies. Comparable results (see Figure 2.10) have been reported in fields other than software.

The extra cost of fixing a late error reflects the cost of locating and undoing the work in all parts or subsystems where the error might have

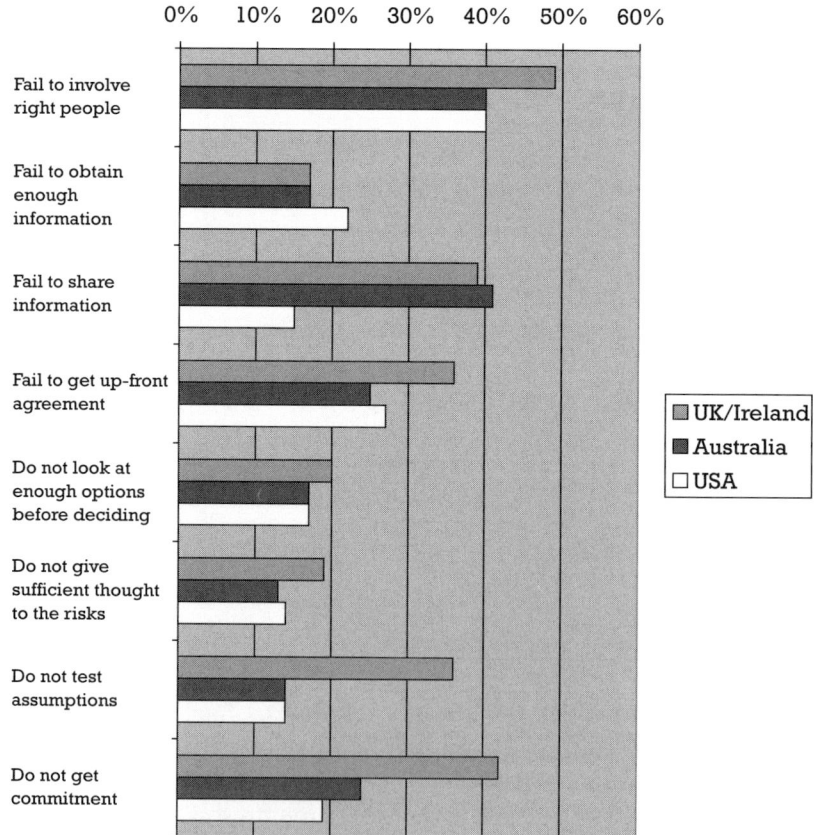

Figure 2.8   Strategies for dealing with time pressure. (*After:* [14].)

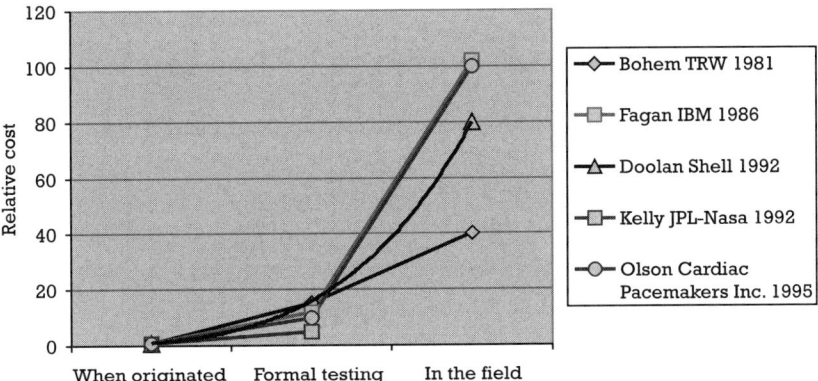

Figure 2.9   Relative costs of fixing problems when originated during formal testing and in the field.

propagated, correcting the original mistake, redoing the work, and finally verifying the correction. So only if the cost of preventing errors were more expensive than the sum of all the other costs would it make sense to eliminate defect-prevention activities in order to recoup project delays.

2.4.4   Overtime

Another typical response to a late project is to "encourage" people to put more hours into the task. We are not talking here about spending a weekend or a late night at the office in order to fix a problem or to prepare an urgent report, but of those projects that require people to put in long hours,

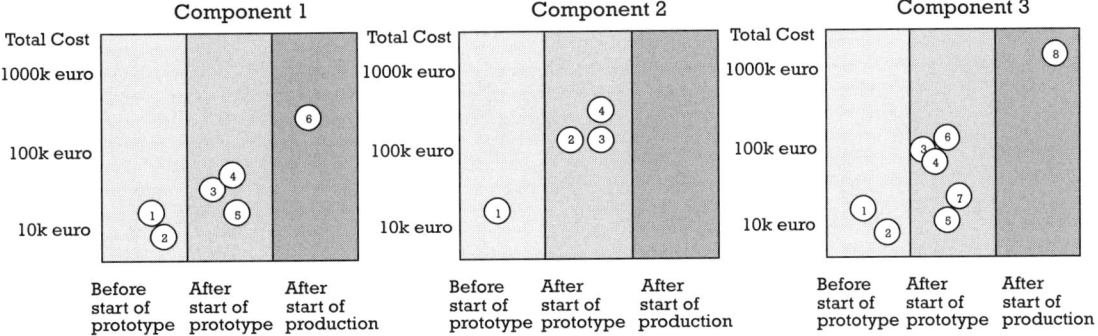

Figure 2.10   Relationship between timing of changes and rework costs in a mechanical engineering project for three different components. Numbers in circles correspond to order of changes. (*After:* [15].)

week after week, weekend after weekend, which has a detrimental effect on their social and private lives. From an organizational standpoint, there are limits to worker productivity no matter how many hours employees put in. In a study conducted by Nevinson [16] that used the results of other such studies, despite the number of hours people stayed at the office, 50 hours of work per week was the maximum in terms of worker productivity (see Figure 2.11). Furthermore, after four consecutive 50-hour weeks, people began to experience burnout, and productivity dropped to less than 35 hours per week.

2.4.5 Fatigue

Tired people not only produce less, they make more mistakes. Human fatigue has long been recognized as a contributing factor in transportation accidents. The Federal Aviation Administration in its analysis of human factors in aviation accidents (see Figure 2.12) attributes 13.6% of accidents due to human error to "adverse mental states" accounting for loss of situational awareness, mental fatigue, circadian dysrhythmia, and pernicious attitudes such as overconfidence, complacency, and misplaced motivation.

In his summary of over 20 years of research studies into operator efficiency as a function of stress and fatigue (see Table 2.6), Hockey [18] reports that fatigue and loss of sleep affect several performance indicators related to the quality of a person's decisions.

The table summarizes the typical outcomes of several studies performed independently over several years. A plus (+) sign indicates an increase in the

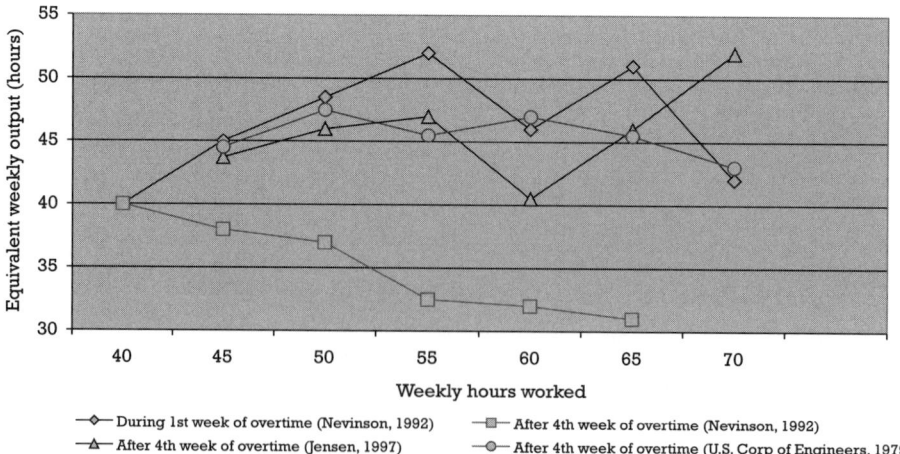

Figure 2.11 Hours at the office versus productive working hours. (*After:* [16].)

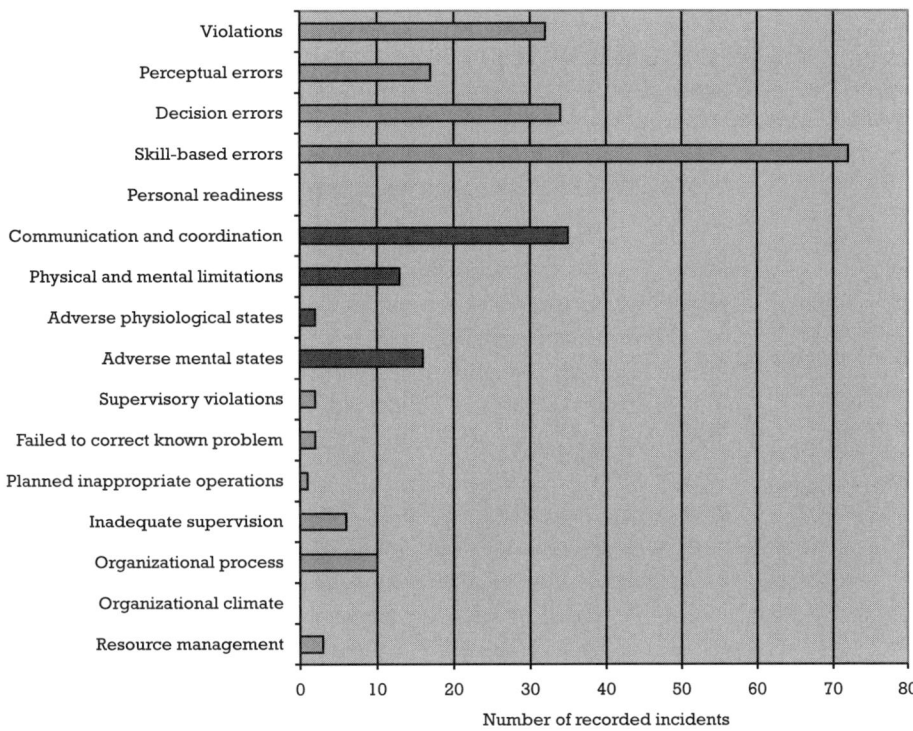

Figure 2.12   Analysis of aviation accidents due to human factors. (*After:* [17].)

Table 2.6   Effects of Fatigue on Performance

Stressors	Performance Indicators				
	Alertness	Attention Breadth	Speed	Accuracy	Short-term Memory Capacity
Noise	+	–	0	–	–
Anxiety	+	–	0	–	–
Incentive	+	–	+	+	+
Stimulant drugs	+	–	+	0	–
Later time of day	+	?	+	–	–
Heat	+	–	0	–	0
Alcohol	–	–	–	–	–
Depressant drugs	–	+	–	–	–
Fatigue	–	–	–	–	0
Sleep loss	–	+	–	–	0
Earlier time of day	–	?	–	+	+

variable measured, a zero (0) either no change or no consistent trend across studies, a minus (–) a decrease in the variable measured. A question mark (?) is used when there is insufficient data to draw conclusions.

2.4.6 Management attention

Figure 2.13 shows the pattern of funds spent, the pattern of funds committed [19], and the pattern of management attention [20] in a typical new-product development project.

What these patterns clearly show is that management attention increases near the final phases of the project, when the wishful thinking, denial, and optimism that reigned over the previous phases must be abandoned, but when there is little possibility of influencing the outcome of the project within original budgeting and schedule constraints. This type of behavior is confirmed by research done by the firm PRTM, which shows that one of the characteristics that distinguishes best-in-class companies is the timing of management intervention, as suggested by the negligible number of projects canceled by these companies in late phases of execution, as compared with project cancellations at other firms (see Figure 2.14).

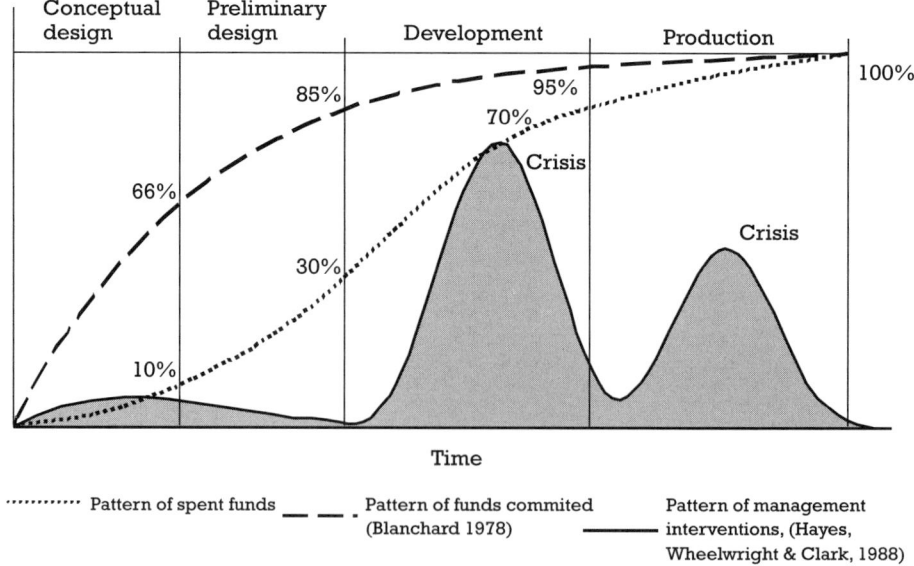

Figure 2.13 Spending, commitment, and management intervention patterns.

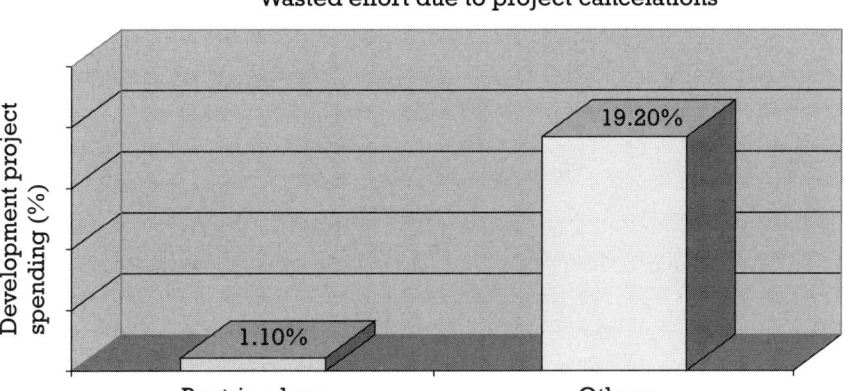

Figure 2.14   Projects canceled after detailed design work in best-in-class firms as compared with other corporations. (*After:* [21].)

2.4.7   Multitasking

As the workload increases beyond the organization's capacity to absorb it, resources tend to be assigned to more than one project at a time in an attempt to keep everything moving and the sponsors happy, but while employees attempt to juggle many activities at once, some "balls" are inevitably dropped. The resulting loss of productivity only makes a bad situation worse. Again, we are not talking here about the sporadic distractions that occur naturally in human endeavors—those involving unexpected problems or personal interactions—but rather the chronic fire-fighting behavior that prevails in many organizations.

Although responsible for some loss of productivity [22] and for some very serious accidents, the multitasking that concerns us is not that which arises from doing many things at once, like having a telephone conversation while reading e-mails or speaking on a mobile phone while driving. I refer here to multitasking that results from individuals being "partially" allocated to several projects at one time. Clark and Wheelwright [20] (see Figure 2.15) detected a sharp drop in value-adding activities as engineers were assigned to more than two projects. This drop is consistent with the results of experiments about learning and forgetting conducted by Spitzer [23] as early as 1939. In these studies Spitzer tested 3,605 students to see how much material was forgotten as a function of time; he discovered that after a day or two, the students had forgotten almost half the material they had read (see Figure 2.16). This loss is even more acute [24] when the subject is simultaneously involved in other learning or problem solving

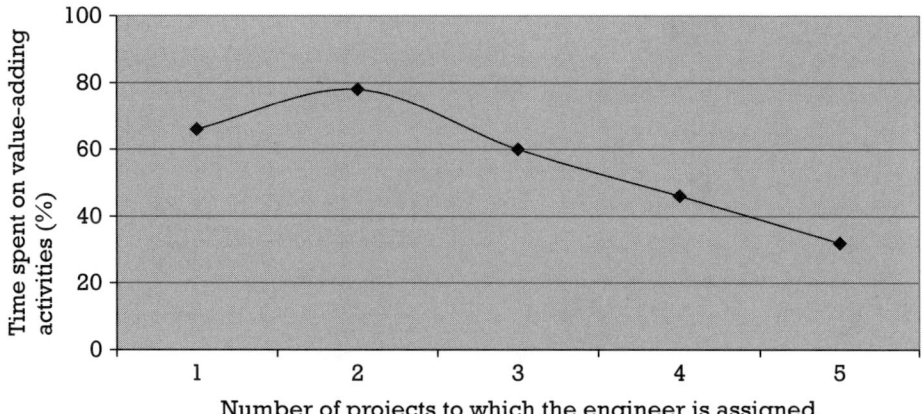

Figure 2.15 Loss of productivity due to task switching. (*After:* [20].)

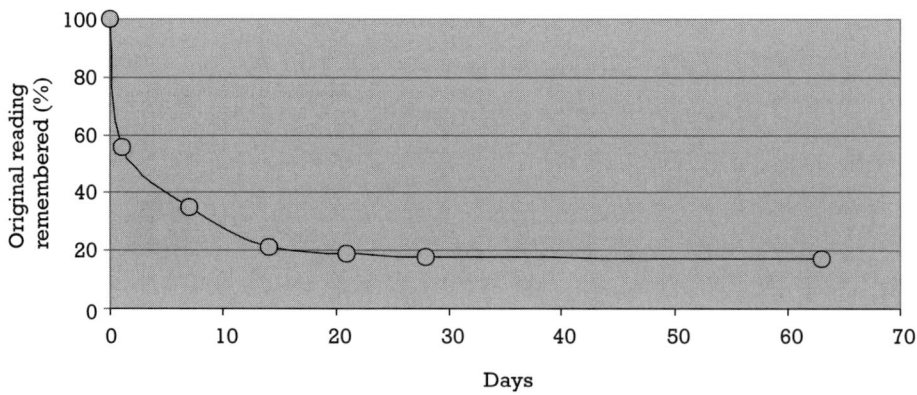

Figure 2.16 Spitzer experiments showing student retention over time. (*After:* [23].)

activities. By drawing a parallel between the students in Spitzer's experiments and an engineer who is interrupted during a design project to work on another task, we can conclude that there is a substantial time requirement in switching tasks, due to the relearning process that must take place when work on the original project is resumed.

But knowledge retention is not the only problem associated with multitasking. If a resource is not ready when required, the associated delay will be passed down the "critical chain" of project activities [25]. Although such disruptions are difficult to quantify in a general way, Figures 2.17 and 2.18

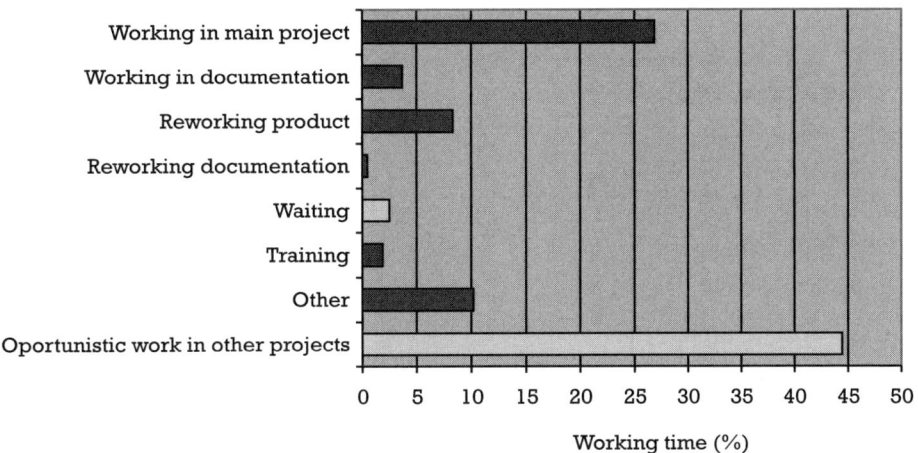

Figure 2.17 Time allocation in project work. (*After:* [26].)

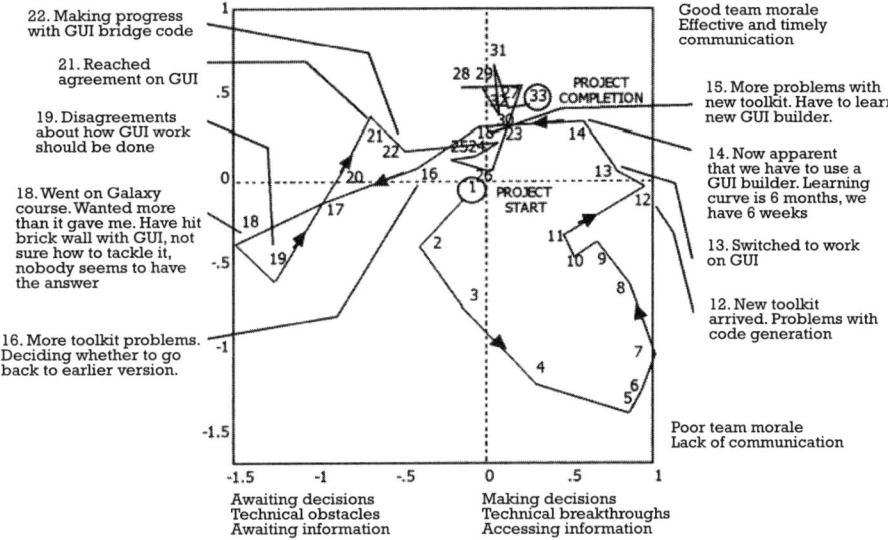

Figure 2.18 Work trajectory for an engineer assigned to a new-product development project. (*Source:* [27].)

show how delays affect an employee's work life, to the extent that some time is inevitably allocated to "waiting" and opportunistic work.

2.4.8 Head count increases

Back in 1975, F. Brooks [28] enunciated his famous law: "Adding people to a late project makes it later." Although today we know that this law should not be applied indiscriminately since the actual outcome of the intervention depends on circumstances such as what stage the project is at, how many people are currently involved, the type of work that must be done, the system architecture, and so on—it is true nonetheless that adding people to a project adds to the existing workload.

First we need to consider the division of work. If the work to be done must be further subdivided to take advantage of the newly added resources, then there will be some added workload to account for the decomposition and later integration of the new parts which were not included in the original plans. Second, the assimilation of the newcomers naturally adds to the workload. During the learning period, which generally ranges from 4 to 8 weeks (see Figure 2.19), not only are the new additions not fully productive, but they take away time from the more senior members of the team, who must coach the newcomers in the intricacies of the project. Third, as the size of the team increases, so does the communication overhead among project members (see Figures 2.20 and 2.21). Larger teams need more time to communicate, achieve consensus, and coordinate their work. This adds to the already overtaxed members of the team, so the immediate consequence of the increase in head count is to slow the project even further. This in turn

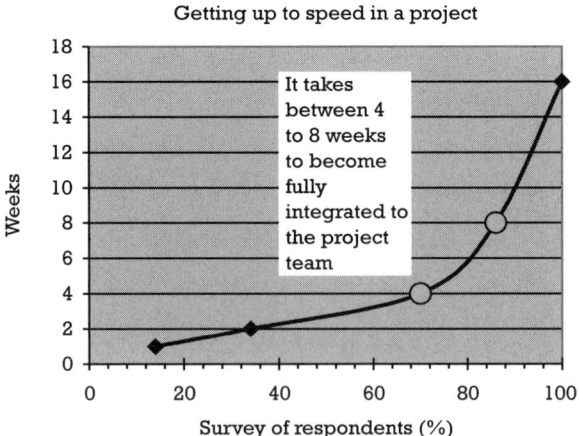

Figure 2.19 According to a majority of Project Managers it takes at least four weeks to get a new team member up to speed. (*After:* [29].)

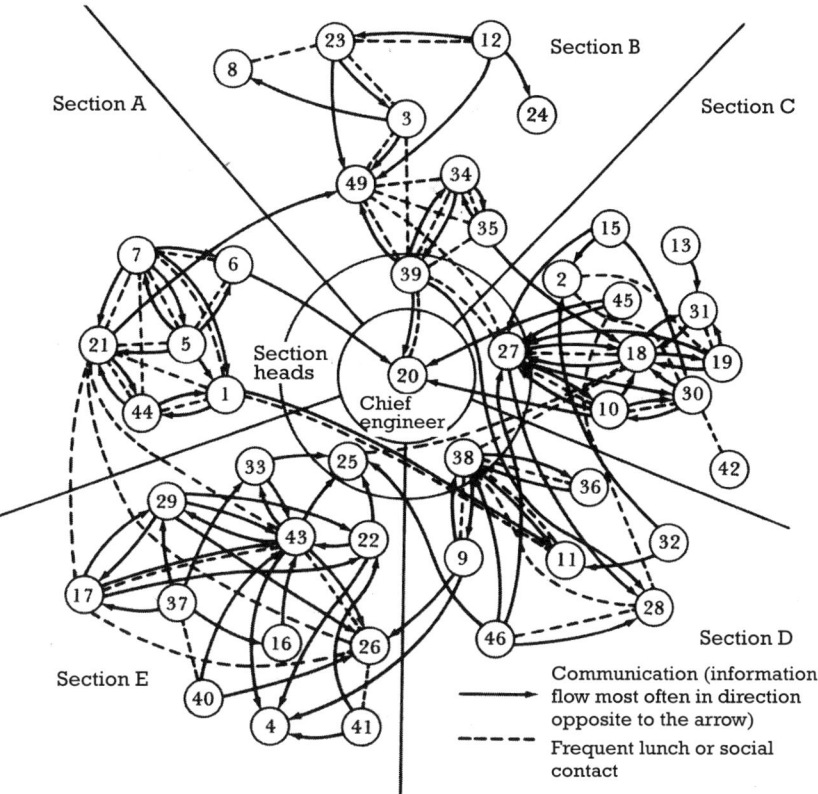

Figure 2.20 Communications path in an organization. Although the original study refers to a functional organization, the same pattern is likely to be found in a project whose sections correspond to subsystems and section heads to team leaders or subproject managers. (*Source:* [30].)

might lead the inexperienced manager to oversteer the project by adding yet more people.

2.4.9 Scope reductions

When all other interventions fail, the only practical thing left to do is to reduce the project's scope. Changing the schedule, the other alternative, is usually not an option, as it is perceived as having too great an impact on the portfolio.

How common is this practice? Figure 2.22 shows the results of a 1995 survey, known as the CHAOS report [321], conducted by the Standish Group, which shows that 53% of the projects were completed over budget,

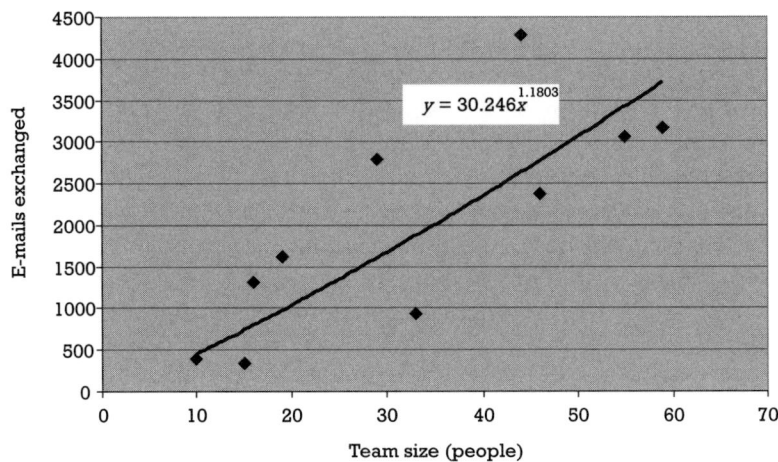

Figure 2.21 Effect of team size in communications. (*After:* [31].)

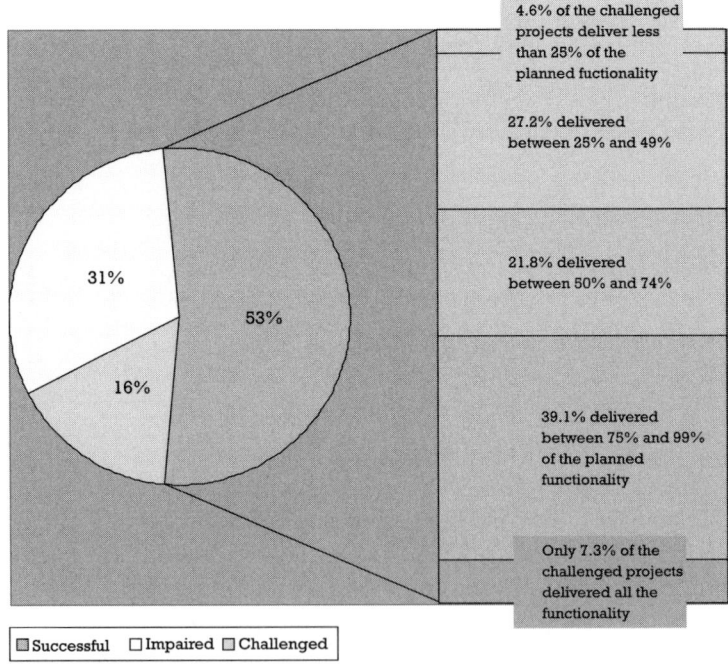

Successful: The project is completed on time and on budget, with all features and functions as
          initially specified.

Challenged: The project is completed and operational but over-budget, over the time estimate, and
          offers fewer features and funcitons than originally specified.

Impaired: The project is canceled at some point during the development cycle.

Figure 2.22 Prevalence of scope reductions to recover from a late project.
(*After:* [32].)

late, and with fewer features than originally specified. Half of this 53% delivered less than 50% percent of the functionality planned.

2.5 Summary

In his book *The Fifth Discipline,* Senge [33] wrote, "Business and other human endeavors are bound by invisible fabrics of interrelated actions, which often take years to fully play out their effects on each other. Since we are part of that lacework ourselves, it's doubly hard to see the whole pattern of change. Instead, we tend to focus on snapshots of isolated parts of the system, and wonder why our deepest problems never seem to get solved." What practitioners commonly call the real world is a mere reflection of this phenomenon. The real world is not the result of natural forces and immutable principles at play; it reflects to a great extent the consequence of past decisions taken without a complete understanding of their ramifications.

A systematic understanding of the limitations of human performance and organizational capabilities is a precondition for meeting the challenges arising from the multiproject environment.

References

[1] Cooper, Kenneth G., "The Rework Cycle: Benchmarks for the Project Manager," *Project Management Journal,* Vol. 24, No. 1, 1993.

[2] Abdel-Hamid, T., and S. E. Madnick, *Software Project Dynamics: An Integrated Approach,* Englewood Cliffs, N.J.: Prentice Hall, 1991.

[3] Sterman, John D., *Business Dynamics: Systems Thinking and Modeling for a Complex World,* Boston: Irwin/McGraw-Hill, 2000.

[4] Browning, T., *Proc. 8th Annual Int. Symp. of INCOSE,* Vancouver, British Columbia, July 26–30, 1998.

[5] Steiner, I. D., *Group Process and Productivity,* New York: Academic Press, 1972.

[6] Repenning, N., P. Gonçalves, and L. Black, *Past the Tipping Point: The Persistence of Firefighting in Product Development,* Cambridge, MA: Sloan School of Management, Massachusetts Institute of Technology, 2001.

[7] Gilliland, S., and R. Landis, "Quality and Quantity Goals in a Complex Decision Task: Strategies and Outcomes," *Journal of Applied Psychology,* Vol. 77, No. 5, 1992, pp. 672–681.

[8] Weinberg, G., and E. Schulman, "Goals and Performance in Computer Programming," *Human Factors,* Vol. 16, No. 1, 1974, pp. 70–77.

[9]   "Process Maturity Profile of the Software Community," Pittsburg: Carnegie Mellon University, Software Engineering Institute, midyear update, 2001.

[10] Mars Climate Orbiter Mishap Investigation Board, Report on Project Management in NASA, NASA, 2000.

[11] Jones, C., "Software Defect-Removal Efficiency," *IEEE Computer*, April 1996.

[12   Kim, J., D. Wilemon, and B. Schultz, "Managing Stress in Product Development Projects," *9th Int. Conf. on Management of Technology*, IAMOT 2000.

[13] "Minds at Work: How Much Brainpower Are We Really Using?" Kepner-Tregoe on-line publication, 1997.

[14] "Decision Making in the Digital Age: Challenges and Responses in the United Kingdom and Republic of Ireland, Australia, and the United States," Kepner-Tregoe on-line publication, 2001.

[15] Terwiesch, C., et al., "Exchanging Preliminary Information in Concurrent Engineering: Alternative Coordination Strategies," Wharton School, 2001.

[16] Nevison, J., "Overtime Hours: The Rule of Fifty," Oak Associates, 1992.

[17] Wiegmann, D., "A Human Error Analysis of Commercial Aviation Accidents Using the Human Factors Analysis and Classification System (HFACS)," Federal Aviation Administration, 2001.

[18] Hockey, G., "Changes in Operator Efficiency as a Function of Environmental Stress, Fatigue and Circadian Rhythms," in *Handbook of Perception and Human Performance*, K. R. Boff, L. Kaufman, and J. P. Thomas (eds.), Vol. 2, *Cognitive Processes and Performance*, New York: Wiley, 1986.

[19] Blanchard, B., "Design and Manage to Life Cycle Cost," Weber Systems, 1978.

[20] Clark, K. B., and S. C. Wheelwright, *Managing New Product and Process Development: Text and Cases*, New York: Free Press, 1993.

[21] PRTM, "Measurement and Management: A Prelude to Action," *Insight*, Vol. 7, No. 2.

[22] Rubinstein, J., D. Meyer, and J. Evans, "Executive Control of Cognitive Processes in Task Switching," *Journal of Experimental Psychology: Human Perception and Performance*, August 2001.

[23] Spitzer, H. F., "Studies in Retention," *Journal of Educational Psychology*, Vol. 30, 1939, pp. 641–656.

[24] Baddeley, A., *Human Memory: Theory and Practice*, rev. ed., Boston: Allyn and Bacon, 1998.

[25] Goldratt, E., *Critical Chain*, Croton-on-Hudson, NY: North River Press, 1997.

[26] Perry, D., et al., "People, Organizations and Process Improvement," *IEEE Software*, July 1994.

[27] Jagodzinski, P., et al., "A Study of Electronics Engineering Design Teams," *Journal of Design Studies*, Vol. 21, 2000, 375–402.

[28] Brooks, F. P., *The Mythical Man-Month: Essays on Software Engineering*, Reading, MA: Addison-Wesley, 1975.

[29] Nevinson, J., "What Can We Learn About Learning in Projects?" *PMNet*, June 1994.

[30] Allen, T., *Managing the Flow of Technology*, Cambridge, MA: MIT Press, 1977.

[31] Bruegge, B., and A. H. Dutoit, Software Metrics for Distributed Development, CMU-CS-96-190, 1996.

[32] The Standish Group, "The Chaos Report," 1994, www.pm2go.com/sample_research.

[33] Senge, P. M., *The Fifth Discipline: The Art and Practice of the Learning Organization*, New York: Doubleday, 1990.

CHAPTER

3

Contents

The project office

In the previous chapter we saw that decisions made in the context of one project can impact other projects in the portfolio in unforeseen and often detrimental ways, and this, coupled with a lack of resource slack, can eventually bring the projects-based organization to a halt. To address these problems, the establishment of a business function responsible for the coordination of all project work across the organization and for providing the infrastructure and competence necessary to manage multiple projects is proposed. We will call this function the project office, or PO.[1]

The PO objective, in contrast with those of a single project, is to complete all projects to best achieve the goals of the organization [11]. The PO's responsibilities include project portfolio management, strategic resource planning, interproject coordination, overall project oversight, cost estimation, contingency planning, quality assurance, external provisioning, project managers' professional development, process management, and tool support.

The PO is an operational function, not a policy-making one. The PO acts as an agent for senior management, providing advice, coordination, and oversight, and although accountable with respect to the execution of the project portfolio, it does not replace either management or the project sponsors with

---

1. This corresponds to the "managerial" type of project office introduced in Chapter 1.

respect to the prioritization of projects and their ultimate disposition.

In this chapter, we will identify the PO's main outputs, its processes and interfaces, and the different competencies or roles necessary to execute them. In subsequent chapters we will address in more detail the process definition, methods, and tools necessary to deploy an effective PO.

3.1 The PO context

A PO can be set up at the business unit level, the product unit level, or at any level at which there arises a need to coordinate multiple projects. Whatever the level within the organizational hierarchy at which the PO is located, it is important that the PO manager has direct access to the same management level as the resource owners. This will help maintain the PO's focus on the interests of the organization as a whole rather than on the interests of any particular functional group, while ensuring that the PO manager has the authority and the access necessary to resolve the conflicts that arise between projects competing for common resources. Figure 3.1 shows the proposed PO reporting relationships. The PO interfaces are shown in Figure 3.2.

Senior management refers to the highest level of management within the organization of which the PO is a part. Senior management is responsible for formulating strategies; it has overall business responsibilities, and

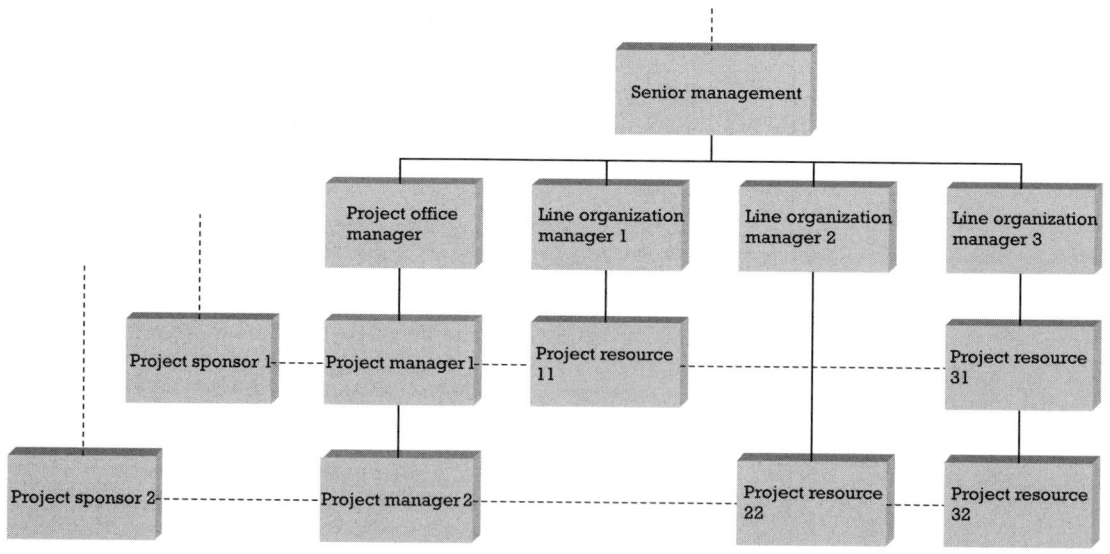

Figure 3.1 PO reporting relationships.

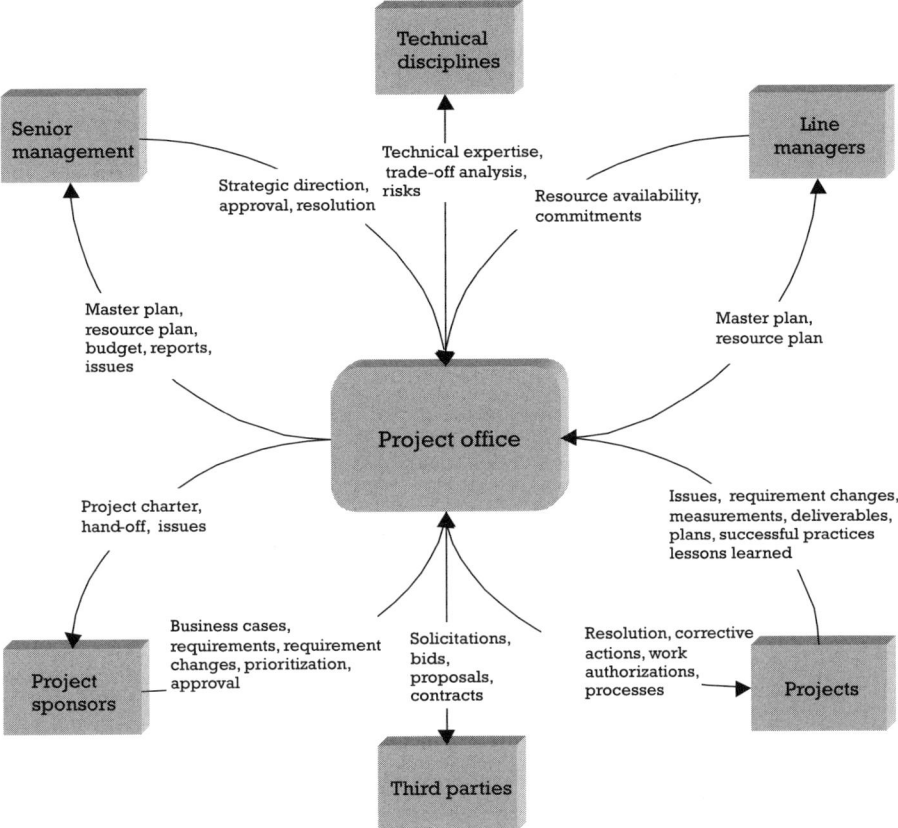

Figure 3.2 PO interfaces.

provides the ultimate decision in the resolution of conflicts. Common titles for senior management are director, vice president, and department head.

Project sponsors are those who request the project work; they have ultimate approval power over expenditures and deliverables. Depending on the business situation, these could be "paying customers," sales representatives, product managers, or any number of internal customers.

Line managers are responsible for the resources to be used in the execution of the projects. They are in general responsible for a function or discipline within the organization. Common titles for line managers are department or section managers.

The technical disciplines entity represents the domain specialists that do not belong to the PO, but who perform work, such as tradeoff studies, on its behalf. These resources usually belong to the line functions.

Third parties are subcontractors, vendors, and other external partners with which the projects are involved in commercial transactions.

In Figure 3.2, the execution of the projects is depicted as external to the PO to emphasize that the day-to-day decisions and the work of the project itself are outside the scope of control of the PO, which intervenes only in case of major deviations and to prevent disruptions to the project portfolio. To do otherwise and involve the PO in every single project decision would result in the establishment of a grinding bureaucracy likely to kill any advantage that might be created by instituting a PO.

3.2   PO information structures

The work of the PO is organized around four fundamental information structures: the master plan, resource plan, financial forecast (see Figure 3.3), and requirements dependency matrix (see Figure 3.4).

The master plan is a time-scaled view of all the projects included in the project portfolio covering a planning horizon of 2 to 3 years. The projects in the plan are portrayed as single tasks characterized by their tentative start dates, their duration, their required effort, their funding needs, and their effort spending profiles. Additional information about the projects could include the degree of commitment to the project (i.e., whether the project is in execution, planned, or envisioned) for those under execution, the status (i.e., whether the project is on time or delayed, and the technologies or products they support). The master plan might also include relationships between projects and links to technology and product road maps.

The resource plan is a forecast, over the planning horizon, of the resources necessary to execute the projects included in the master plan. The resource plan covers the current availability of resources (head count), their competencies, a recruiting plan, and periods during which excess capacity might exist. The resource plan shows whether the resource utilization is based on current, planned, or envisioned work. At this level, the resource plan is prepared based on the competence of the resources and not by assigning specific individuals to the projects. Plans for resources such as test benches, laboratories, and computing equipment are better taken care of by the line organizations that own them.

The financial forecast depicts the cash flows, expenses, and revenues arising from the execution of the projects in the master plan, with the purpose of helping senior management and project sponsors choose the portfolio configuration that best meets the objectives and capabilities of the

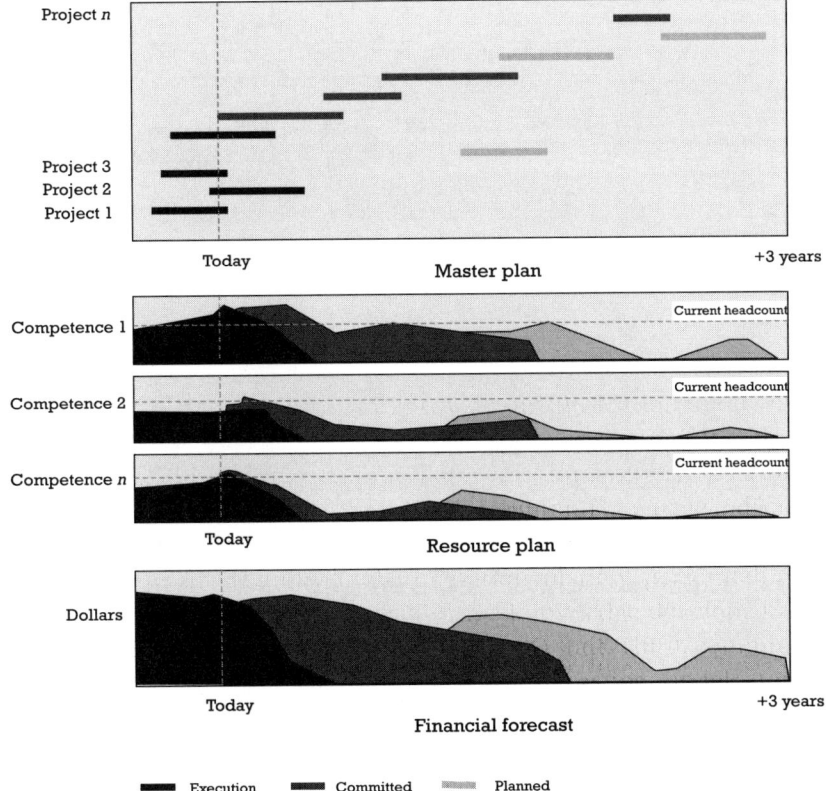

Figure 3.3   Master plan, resource plan, and financial forecast.

organization. The financial information contained in the forecast includes labor costs, nonlabor costs, management reserves, volume allowances, and funding sources. In addition to the financial forecast, the PO also prepares detailed quarterly or annual budgets for the projects in execution and for those beginning in the next budgeting period.

The requirements dependency matrix is an important tool for organizations working on product lines or whose products evolve through successive reincarnations of added functionality. The matrix links the requirements or features to be developed in future projects to those in previous projects that will serve as a foundation upon which the latest will be built. The matrix allows tracing the consequences of postponing or canceling the implementation of any feature through the entire project portfolio. Additionally, the matrix might contain financial and effort information that allows calculating the impact—in terms of lost revenues and extra development effort—that

Projects	Where used	Project 1	Project 1	Project 2	Project 2	...	Project 2	...	Project n
Where developed	Requirements	Feature 1-1	Feature 1-2	Feature 2-1	Feature 2-2		Feature 2-n		Feature n-1
Project 1	Feature 1-1			▲		▲			▲
Project 1	Feature 1-2			▲			▲		
Project 2	Feature 2-1								
Project 2	Feature 2-2								
⋮							▲		
Project 2	Feature 2-n								▲
⋮									
Project n	Feature n-1								

Feature n-1 in Project n requires the implementation of Features 1-1 and 2-n by projects 1 and 2 respectively. Feature 2-n requires feature 1-2 to be developed by Project 1. Therefore, if Feature 1-2 is dropped from Project 1, Project 2, and Project n will be impacted.

Figure 3.4   Requirements dependency matrix.

such decisions would have over subsequent projects. The mechanism for this will be explained in detail in Chapter 6.

3.3   PO processes

The processes performed by a managerial type of PO revolve around three main themes:

1.   Project life-cycle management;

2.   Project portfolio management;

3.   Support functions.

The most important of these, the one that provides the justification for a heavyweight PO like the one proposed here, is the second item, project portfolio management, shown at the center of Figure 3.5.[2] If the organization does not adhere to the portfolio concept or if the portfolio is small, then probably all that the organization needs is a "repository" or "coach" type of

---

2. The PO processes have been described using IDEF0 notation. See Appendix A for an explanation of this notation.

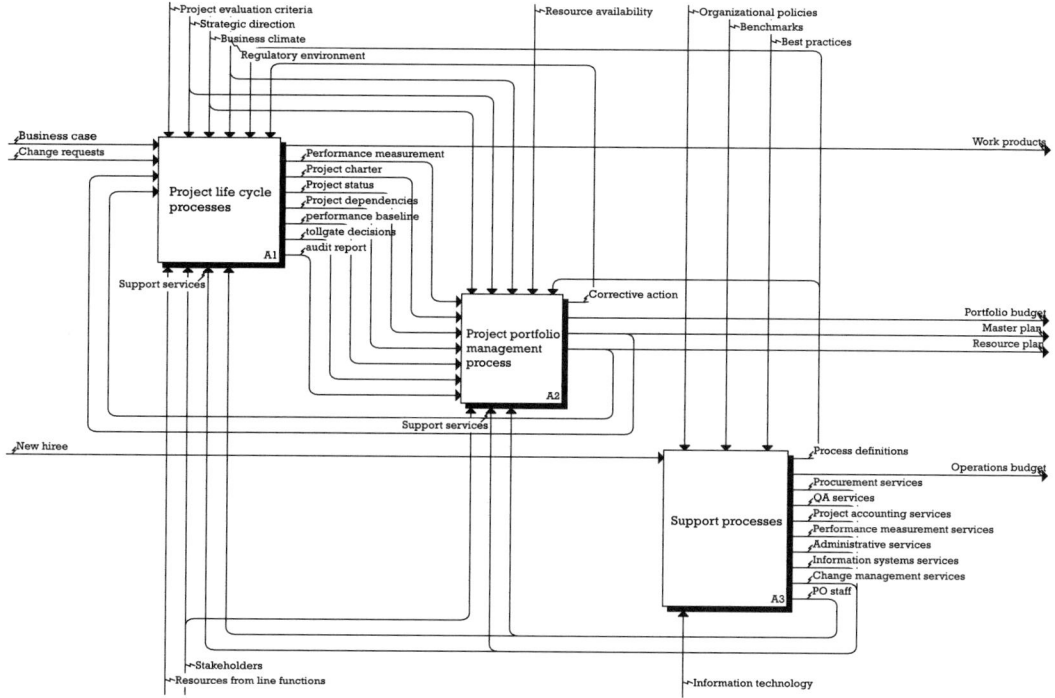

Figure 3.5 PO main processes.

PO (see Section 1.3), with the rest of the responsibilities shared between senior management and line functions.

In deciding which processes to include under the responsibility of the PO, there are two important criteria to consider: First, the PO should be accountable and have authority over those processes that clearly fall under its area of responsibility, such as portfolio and project management. Second, the PO should have responsibility over those processes that allow it to stay "in the loop." The justification for the second requirement is simple: In order to assure that the PO has the necessary power to exercise its authority effectively and that it is not bypassed when important decisions need to be made, the PO must have a hand in such processes as change management, vendor selection, career and professional development for project managers, and budgeting.

Although the PO must be given responsibility for the execution of these functions, the work to be performed is not limited to the PO staff. For example, formulating a project charter would require the involvement of personnel from the sponsoring organization as well as specialists borrowed from the different technical disciplines. This is necessary not only because the PO does not have all the technical resources needed to accomplish the task, but because such involvement helps foster consensus among the stakeholders. Another example is the portfolio planning process, where the PO manager acts as a convener and facilitator, with the final decisions taken by senior management in conjunction with the project sponsors.

3.3.1 Project life-cycle process

The project life-cycle process addresses the formulation, planning, execution termination, and review of individual projects (see Figure 3.6).

3.3.1.1 Project formulation

Upon receiving a request for a new project or a major change to an existing one, the PO conducts a preliminary study to establish its scope, work approach, duration, effort required, and other business aspects. The extent of the work to be performed at this point is limited to that necessary to make an informed decision with regards to whether or not to include the request in the portfolio mix (see Figure 3.7).

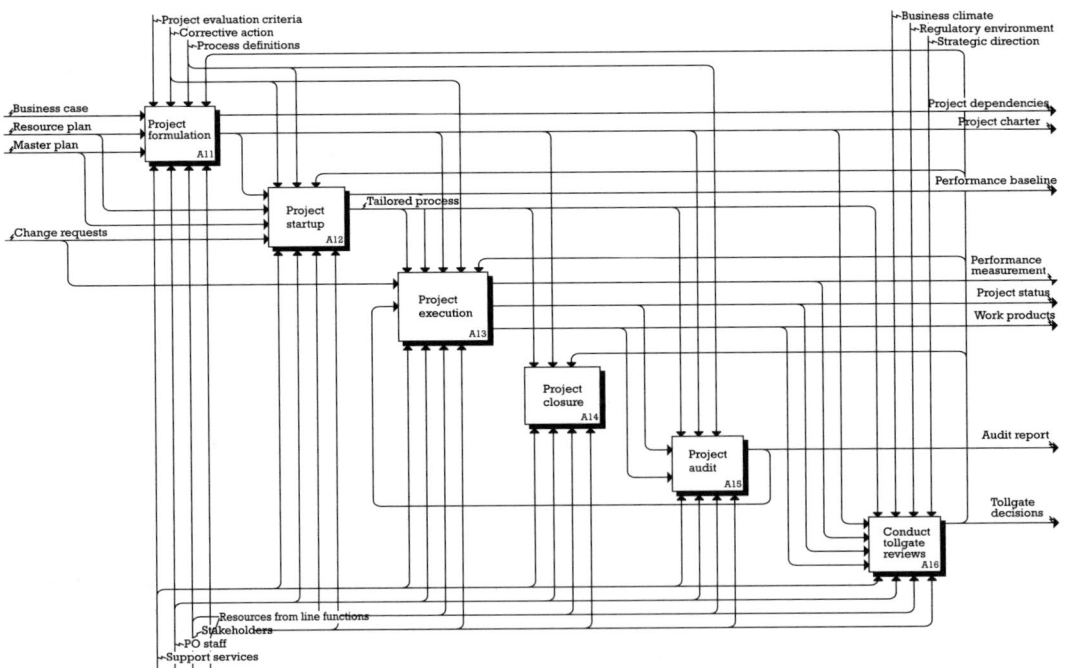

Figure 3.6 Project life cycel processes.

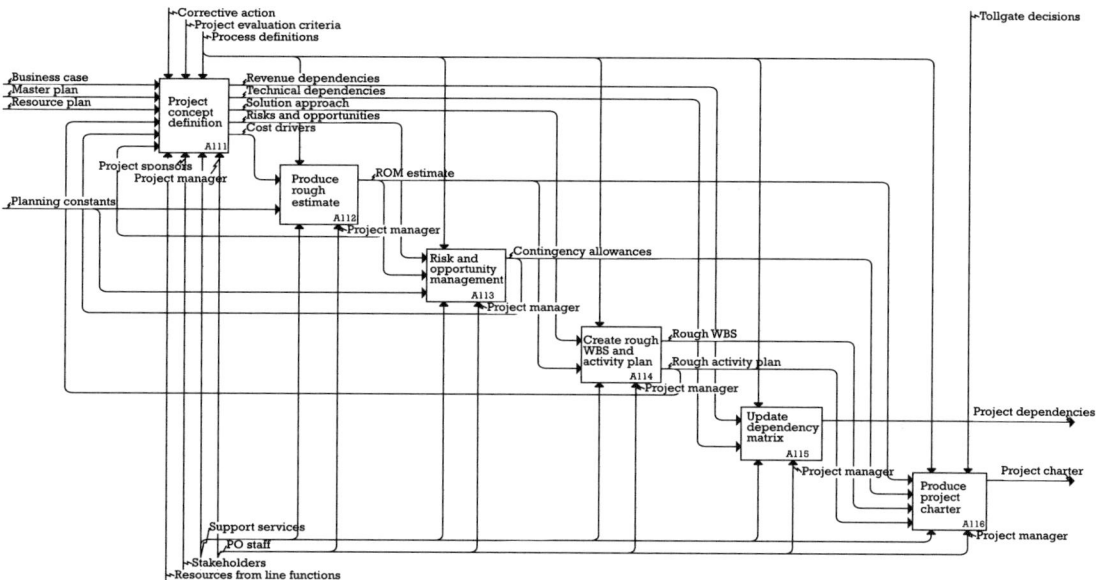

Figure 3.7 Project formulation process.

To support this activity, the PO will typically set up a multidisciplinary team with the participation of specialists from various departments and representatives of the sponsor. The main output of this activity is a project charter, which specifies the scope of work, the time frame in which the work is to be performed, the effort and other resources necessary for the execution, the major risks that could derail the project, and links to other projects in the portfolio. The project charter will be refined as work progresses. Contingency funds are evaluated at this time, with the purpose of minimizing costs by spreading the risks across all projects, much in the way that an insurance company will do with respect to its policyholders.

3.3.1.2 Project startup

The project core team is assembled. The project scope, the initial estimates, the assumptions, and the work approach proposed during the project formulation phase are revisited. Resource coverage is verified and necessary changes agreed to with the project sponsor. Changes that might have an impact on other projects are submitted for review and approval in the context of the master plan. The WBS is refined, work packages defined, and cost accounts set up. A performance baseline, against which performance will be measured, is established. Project staffing and work is then begun according to the project plan (see Figure 3.8).

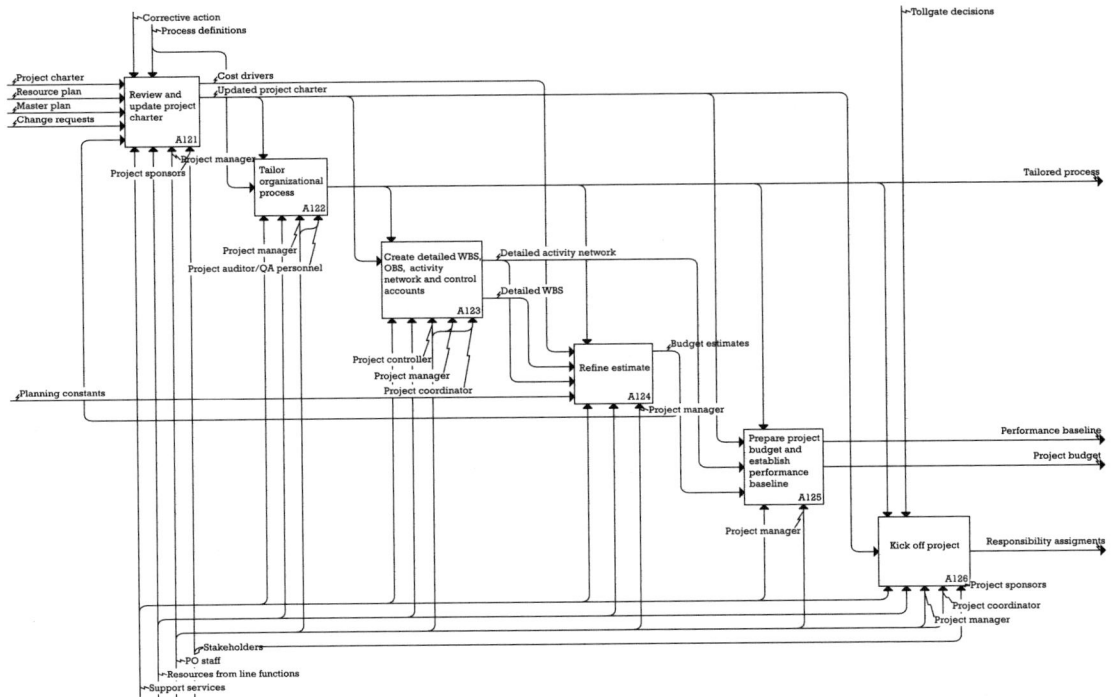

Figure 3.8 Project startup process.

It is during this phase that the organizational project management method and other processes are adapted or tailored to the circumstances and needs of the project.

3.3.1.3 Project execution
It is at the project-execution stage (see Figure 3.9) that the actual project work gets done. The project-execution process brings together, into a temporary organization, the resources belonging to different line functions to work in a common endeavor, in accordance with the specifications contained in the project charter. The responsibility of the PO is exercised through the project manager, who is responsible for producing the desired results on time and within budget, for encouraging teamwork and commitment, and for ensuring that the processes, methods, and standards of the larger organization are adhered to.

3.3.1.4 Project closure
As the project draws to a close, the PO must ensure that all the work is completed, that the people finalizing their assignments are recognized for their

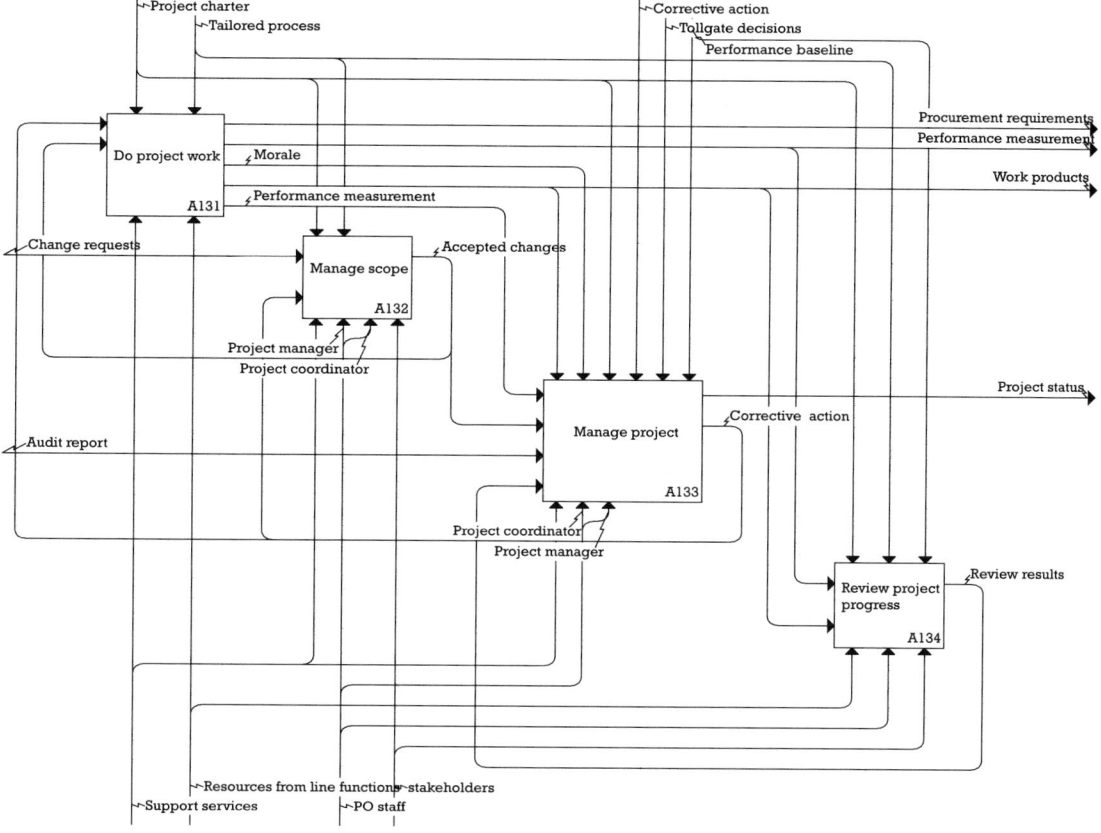

Figure 3.9   Project execution process.

contributions, and that the lessons learned are incorporated into the project-management processes.

The key activities at this stage are work completion, transferring owner-ship of the deliverables, closing contracts with subcontractors and suppliers, debriefing the project team, conducting a lessons-learned exercise, reward-ing achievement, and disbanding the project team (see Figure 3.10).

3.3.1.5   Project audit

A project audit is an in-depth evaluation of the "true and fair" state of a project conducted by a person not belonging to the project team. A project audit has as its purpose one or more of the following:

- ▸ To ensure that the work is being performed in accordance with estab-lished procedures;

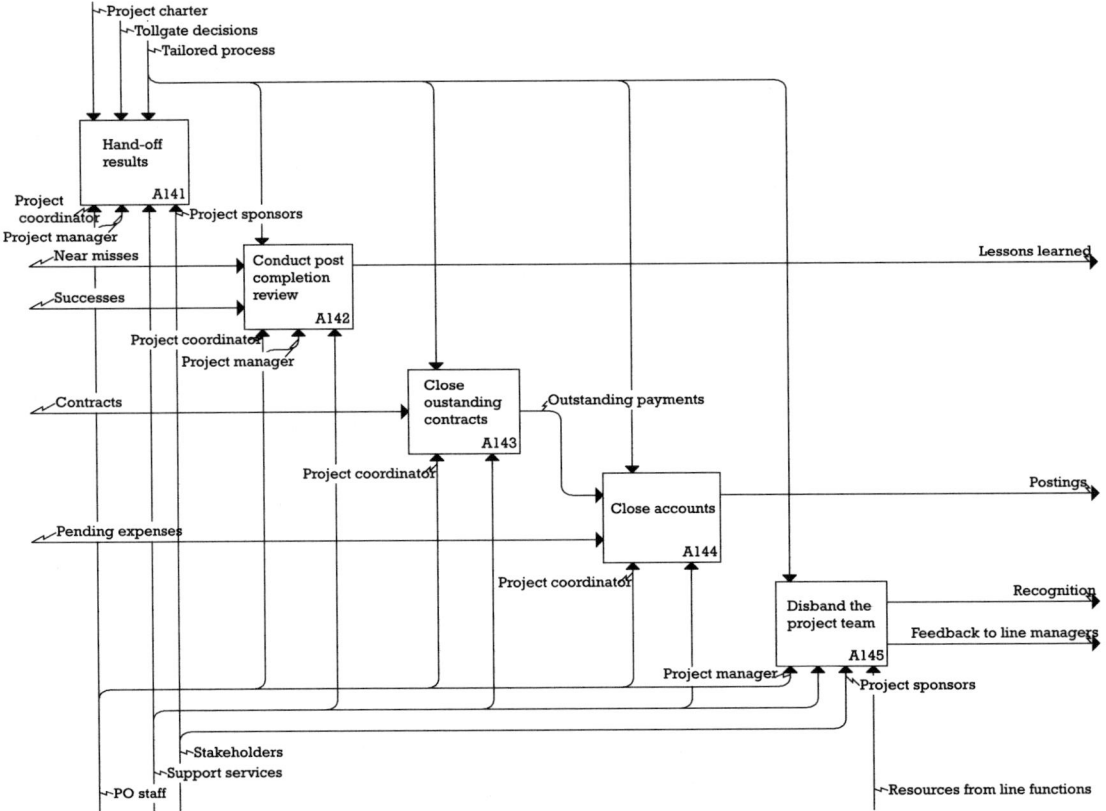

Figure 3.10 Project closure process.

▸ To establish the real condition of a project in terms of time, quality, cost, scope, customer satisfaction, and employee moral.

In order to gather data for the audit, the assessors will check work procedures, documentation, and deliverables; conduct interviews with customers, employees, and suppliers; and conduct a root-cause analysis. The findings of the audit are then documented in an audit report for management follow-up (see Figure 3.11).

3.3.1.6 Tollgate reviews

Tollgates are standard decision points in the life of the project. At each tollgate, a decision is made on whether to continue with the project, to kill it, or to change it in some significant way (see Figure 3.12).

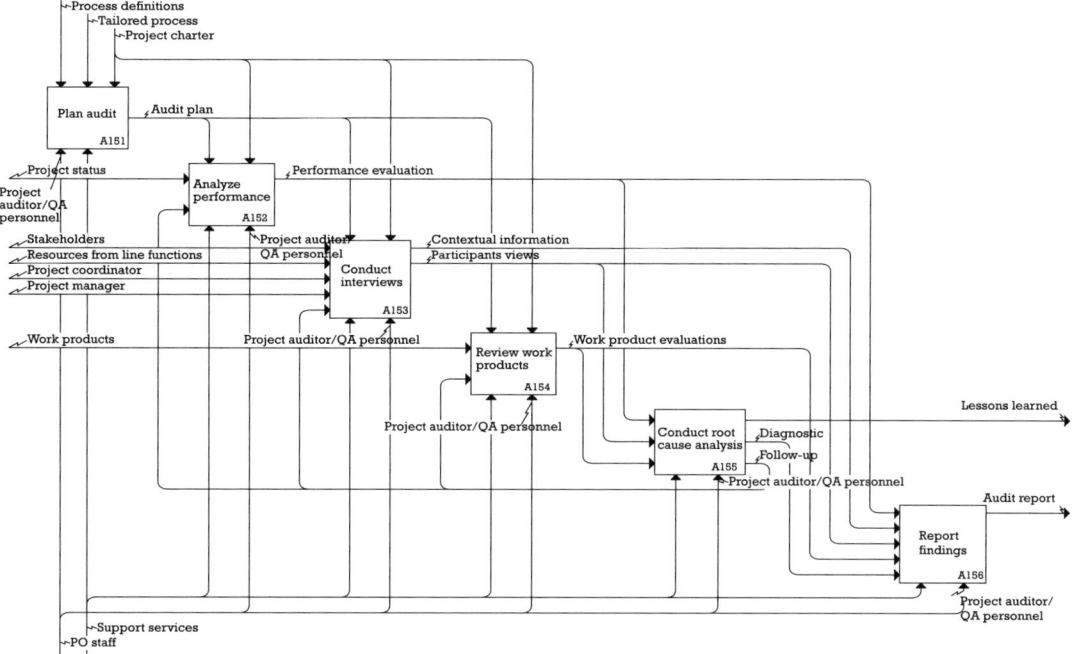

Figure 3.11   Project audit process.

At each tollgate, a project will be reviewed from at least three different perspectives: business, progress, and cost. Representative questions are as follows: Are the justifications for this project still valid? Is the project making progress as expected? Are solutions appearing faster than problems, or vice versa? Are resources being used efficiently? What will be the cost at completion?

An effective tollgate process, one with the ability to discriminate between bad and good projects and a willingness to terminate the losers, is a key factor distinguishing "best-in-class" organizations from the rest.

3.3.2   Portfolio-management process

The portfolio-management process comprises portfolio planning, project oversight, and portfolio control (see Figure 3.13). The portfolio management process seeks to maximize the benefits that can be attained, with a given level of risk, from all of the projects currently undertaken by the organization and those envisioned for the years to come.

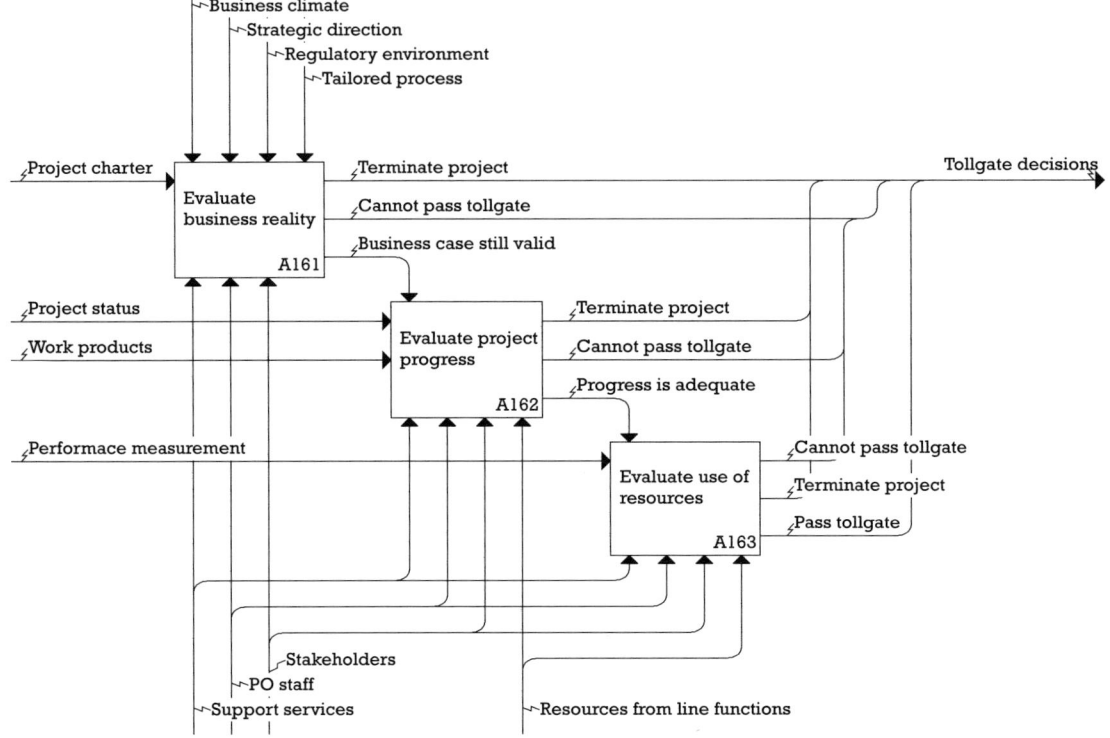

Figure 3.12   Tollgate process.

3.3.2.1   Project portfolio planning

Project portfolio planning is the point at which projects and business come together. The outcome of this process is a plan that balances work, results, resources, and risk according to the objectives of the organization. It involves deciding which projects to execute and when, forecasting the resources needed to execute the selected projects, and projecting the resulting cash flows.

The project-portfolio-planning process (see Figure 3.14) is performed at regular intervals, usually quarterly, or when special circumstances such as major deviations in individual projects, reorganizations, or new opportunities impose a revision of existing plans. At this level, projects are viewed as a single task. The selection and prioritization of projects is made with the objective of striking a balance between criteria such as the projects' strategic position, probability of technical success, probability of commercial success, sociopolitical and regulatory consequences, costs, rewards, stage of

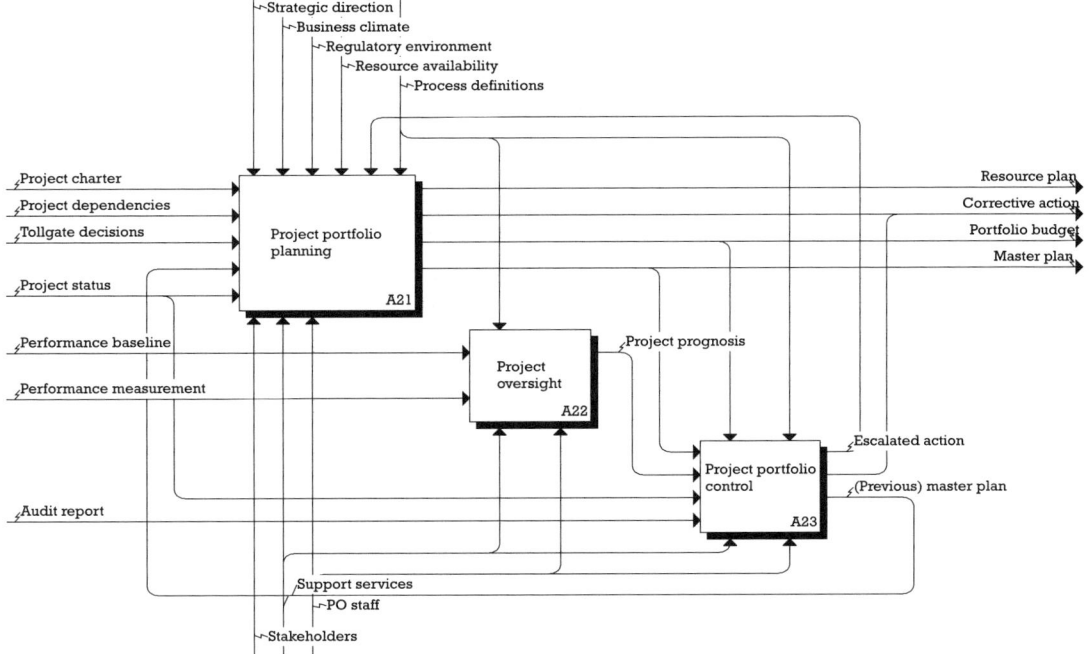

Figure 3.13   Portfolio-management process.

innovation, and resource constraints. Specific techniques for balancing the project portfolio will be dealt with in Chapter 6.

3.3.2.2   Project oversight

The purpose of project oversight is to provide early warnings about the performance of individual projects so that management can act before local issues spread to the entire portfolio.

To provide early warnings (see Figure 3.15), the project's performance must be assessed against its performance baseline and against the output of forecasting models built out of measurements collected from previous endeavors. The output of the process is a prognosis of the project's health, assessed by doing "checkups" on the following areas: progress, cost, quality, and staff morale. Specific quantitative techniques will be provided in detail in Chapter 7.

Quantitative oversight is not a substitute for, but a complement to, direct observation. The project manager and the PO manager must walk around to corroborate, in the field, whatever the indicators might be saying and to pick up signals overlooked by the current process.

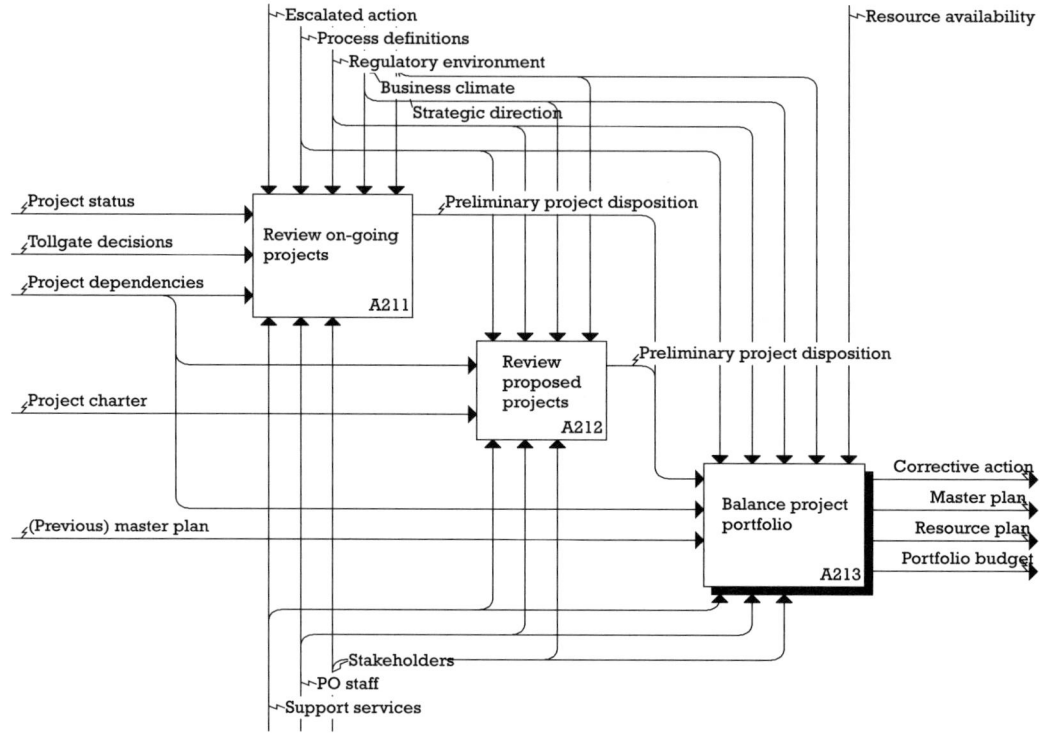

Figure 3.14 Project portfolio-planning process.

3.3.2.3 Portfolio control

Portfolio control is the process by which the PO takes action to compensate or minimize the impact of project deviations over the entire portfolio.

Estimates are reviewed to ascertain that all projects will be completed in the allocated time frames, that resources will be freed on time, and that the impacts of cross-project delays are minimized. Appropriate corrective actions are decided in the context of the master and resource plans and not on the basis of the affected project alone (see Figure 3.16).

The PO manager will take action by rebalancing the portfolio within the time-resource window defined for each project. Beyond those parameters, conflicts would need to be referred to senior management for resolution.

3.3.3 Support processes

The support processes provide the foundation on top of which all the other processes operate (see Figure 3.17). Despite their low visibility, these

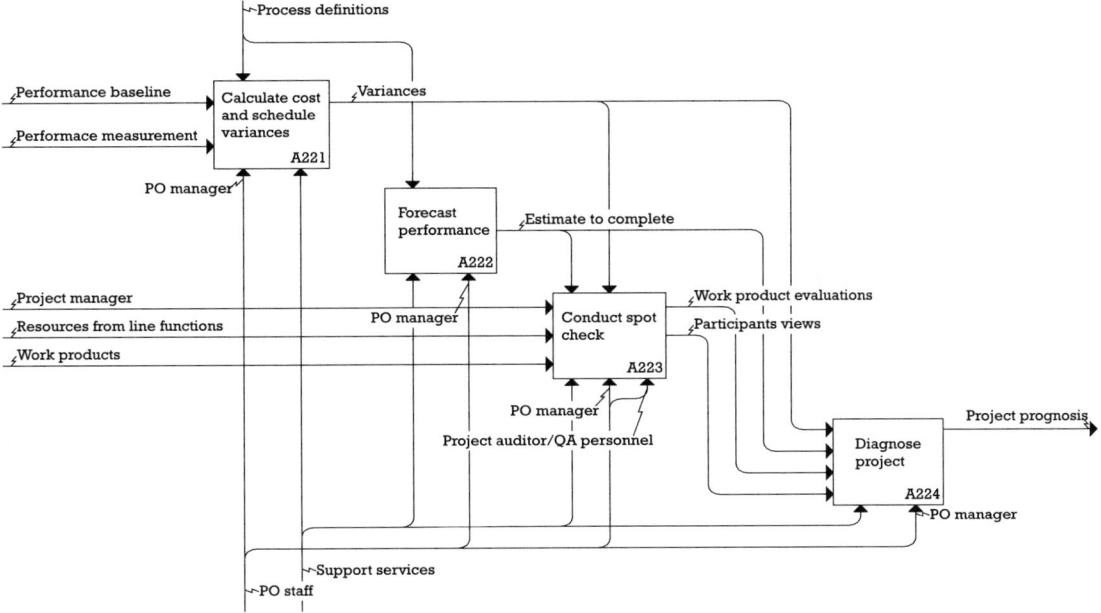

Figure 3.15 Project oversight process.

processes are an essential component of the PO. Furthermore, in the case of the PO as repository or coach, the support processes are the only processes specifically assigned to it.

The importance of these processes resides in the fact that it is through them that the PO can ascertain project and activity progress without interfering with the project work. For example, by examining the activity logs of the configuration management system it is possible to determine the status of the work in process or the number of changes requested by a sponsor—of course, the system must first be designed to provide this information. Similarly, if the training and evaluation of project managers were not under the control of the PO, it would be very difficult for the PO manager to exercise authority over them.

There are eight fundamental support processes:

1. Processes and information systems management;

2. Measurement process;

3. Change management;

4. Procurement management;

5. Quality assurance;

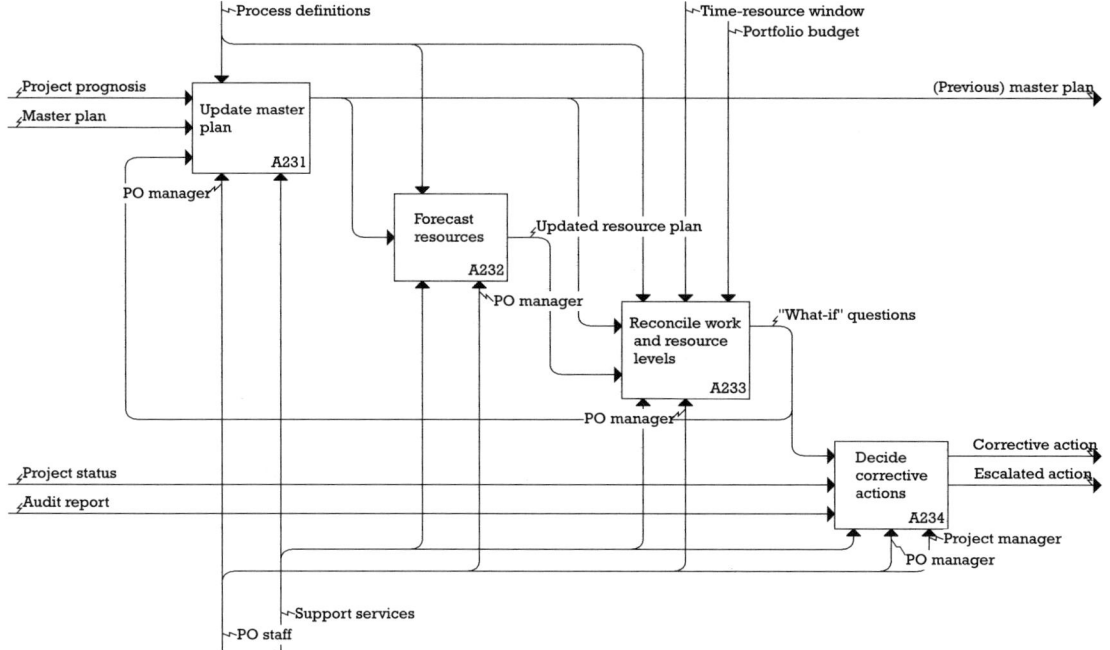

Figure 3.16 Portfolio control process.

6. Administrative support;

7. Project accounting;

8. Human-resources management.

3.3.3.1 Processes and information systems management

The success of the PO relies on the existence of common processes and tools. Without them, the system is unmanageable. But as important as the role that processes and tools play in developing a common vocabulary is their value as intellectual capital and as a source of competitive advantage. Processes and tools are the embodiment of the collective knowledge developed by the organization.

The notion of process improvement embraced here is based on the notion of bottlenecks [2]. Bottlenecks are activities or mechanisms that limit the throughput of systems along a given dimension: time to market, quality, and so on. Improvements in areas other than the bottleneck do not result in a performance increase at the system level. Of course, once we have removed a given constraint, a bottleneck will appear elsewhere and the process will be repeated. By focusing the improvement work where it

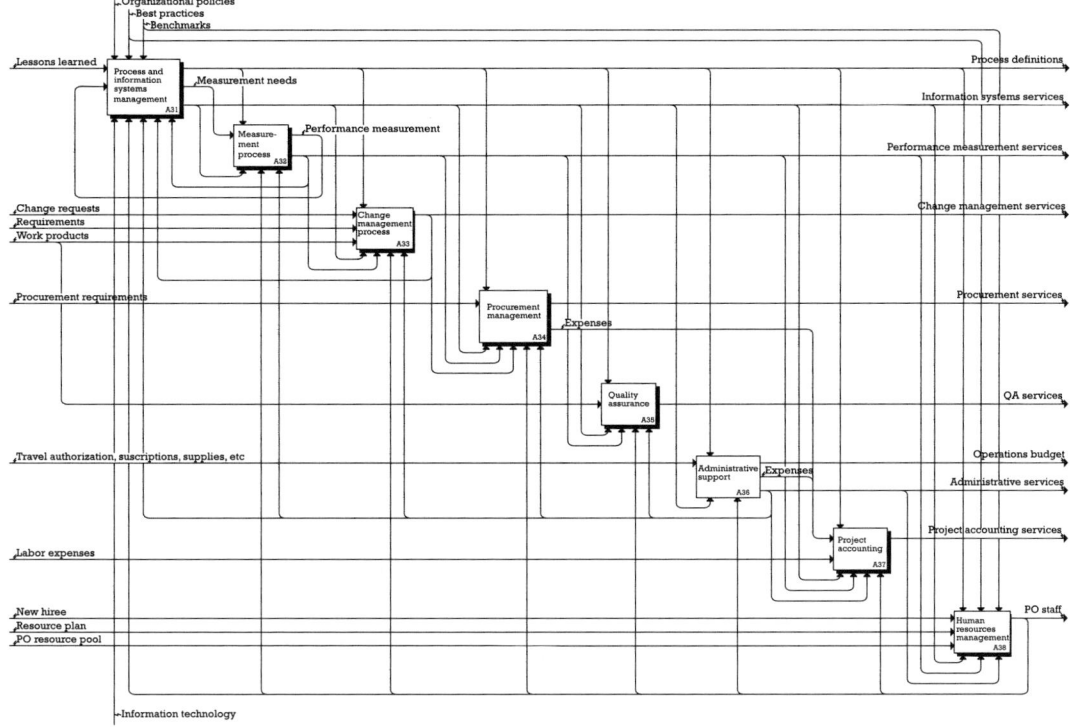

Figure 3.17 Support processes.

really matters, not only do we reduce cost but we also minimize disturbance to the ongoing work, which in turn results in less variability. So by improving the improvement process we could achieve an improvement in the overall process.

Processes and tools are improved based on information coming from industry, academia, experience gained from the execution of projects, and from the insights of PO personnel (see Figure 3.18).

3.3.3.2 Measurement process
The measurement process involves three activities (see Figure 3.19):

1. Planning the measurements;

2. Performing the measurements;

3. Producing performance statistics.

A detailed explanation of this process is given in Chapter 7.

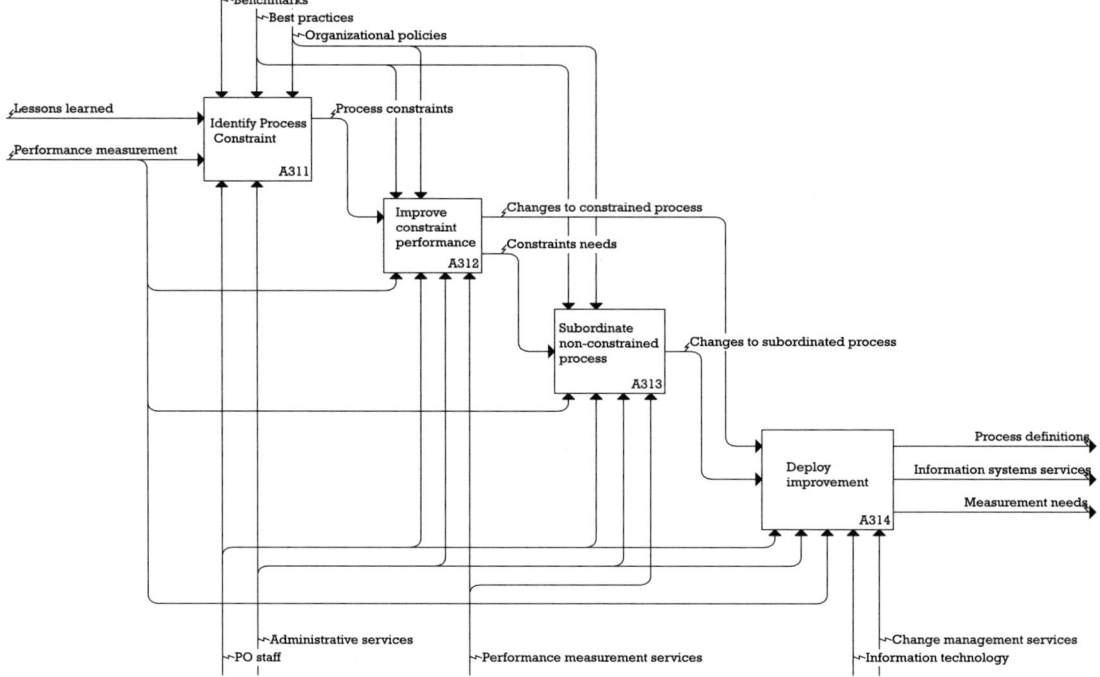

Figure 3.18 Process and information systems management.

3.3.3.3 Change management process

Change and project work are inseparable. Change occurs naturally as part of the work that is done within the project, in response to changes in the business environment and to changes in the wishes and needs of the project sponsors.

Change management is a pervasive process that touches on every aspect of the project work. With respect to the project sponsor, it deals with changes to the project scope; within the project it deals with the evolution of the project's work products. Simply stated, the purpose of change management is to maintain in a congruent state plans, contracts, requirements, and specifications.

Change management involves three interrelated efforts (see Figure 3.20):

1. *Requirements management:* The purpose of the requirements management process is to manage the requirements of the project's products and product components and to identify inconsistencies between those requirements and the project's plans and work products. The

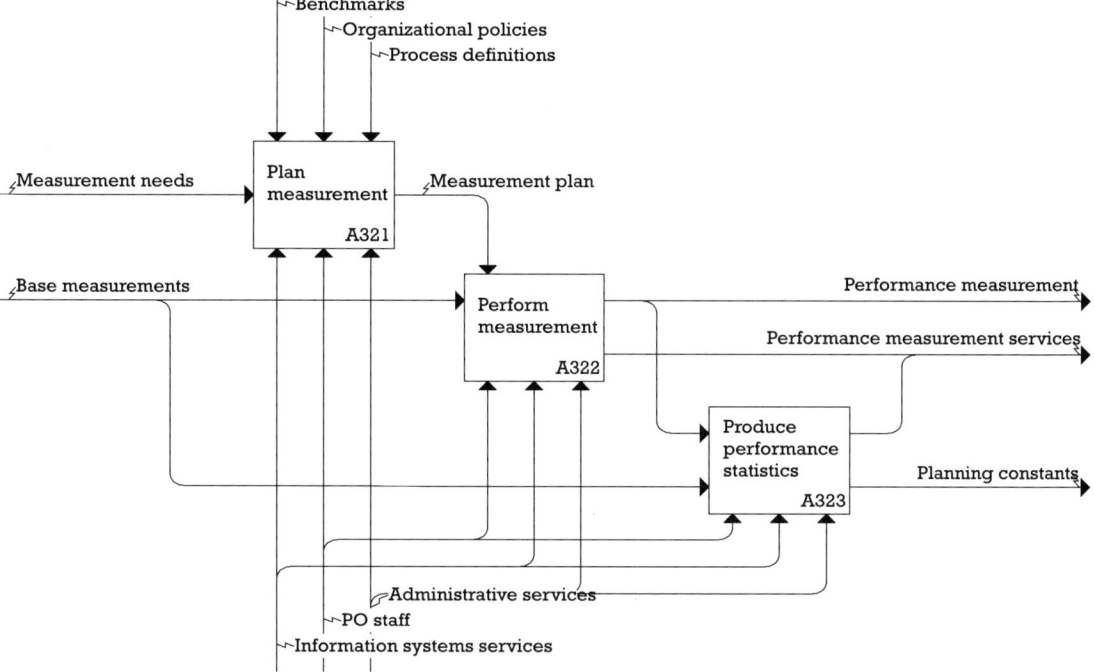

Figure 3.19 Measurement process.

management of requirements involves documenting requirements changes and rationales and maintaining bidirectional traceability between source requirements and all product and product-component requirements [3].

2. *Configuration management (CM):* The purpose of the CM process is to establish and maintain the integrity of project work prod-ucts—including products that are delivered to the customer, designated internal work products, acquired products, tools, and other items that are used in creating and describing these work prod-ucts—and of organization work products, such as standards, procedures, and reuse libraries. The CM process involves identifying the configuration of selected work products that compose the base-lines at given points in time, controlling changes to configuration items, building or providing specifications to build work products from the configuration management system, maintaining the integ-rity of baselines, and providing accurate status and current configuration data to developers, end users, and customers [3].

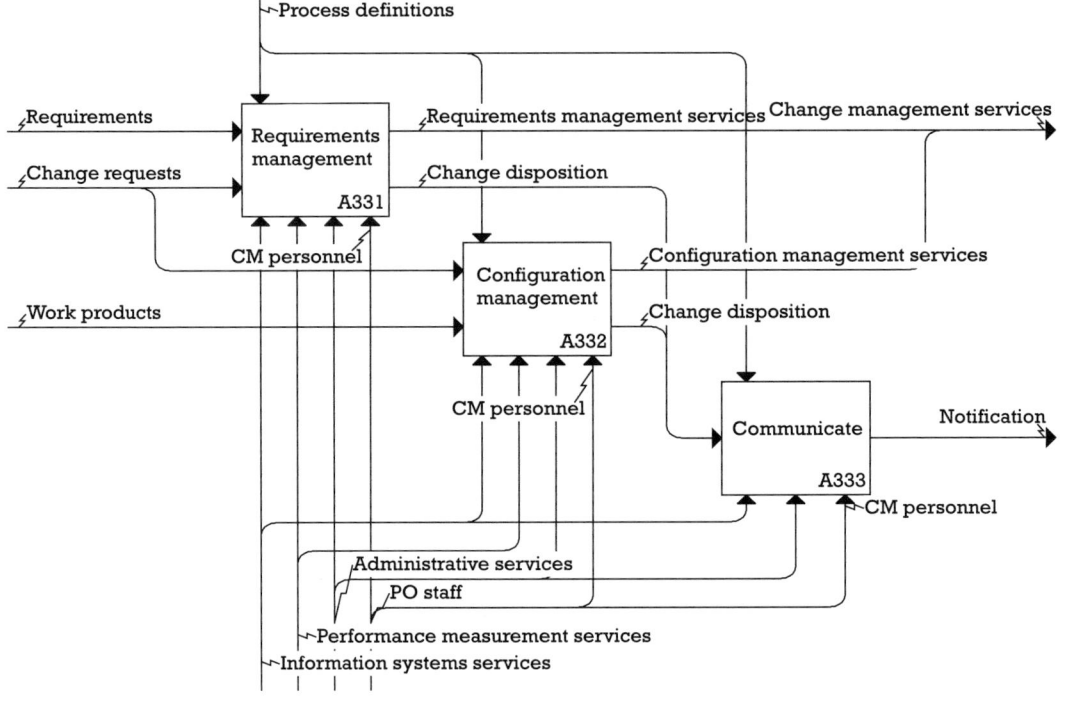

Figure 3.20 Change management process.

3. *Communications:* The purpose of this effort is to ensure that all parties are informed of the disposition and consequences of proposed changes.

3.3.3.4 Procurement management
The purpose of this activity is to support project managers in dealing with third parties, vendors, and subcontractors involved in their projects.

Procurement management involves choosing the acquisition strategy, the selection of suppliers, the negotiation of contracts, and the tracking and auditing of third-party capabilities, performance, and results (see Figure 3.21).

3.3.3.5 Quality assurance
The quality assurance (QA) process concerns the periodic check of work products and work processes employed by the projects and by the PO. Do we do what we say? Do we observe our own procedures? Do we keep the documentation up to date?

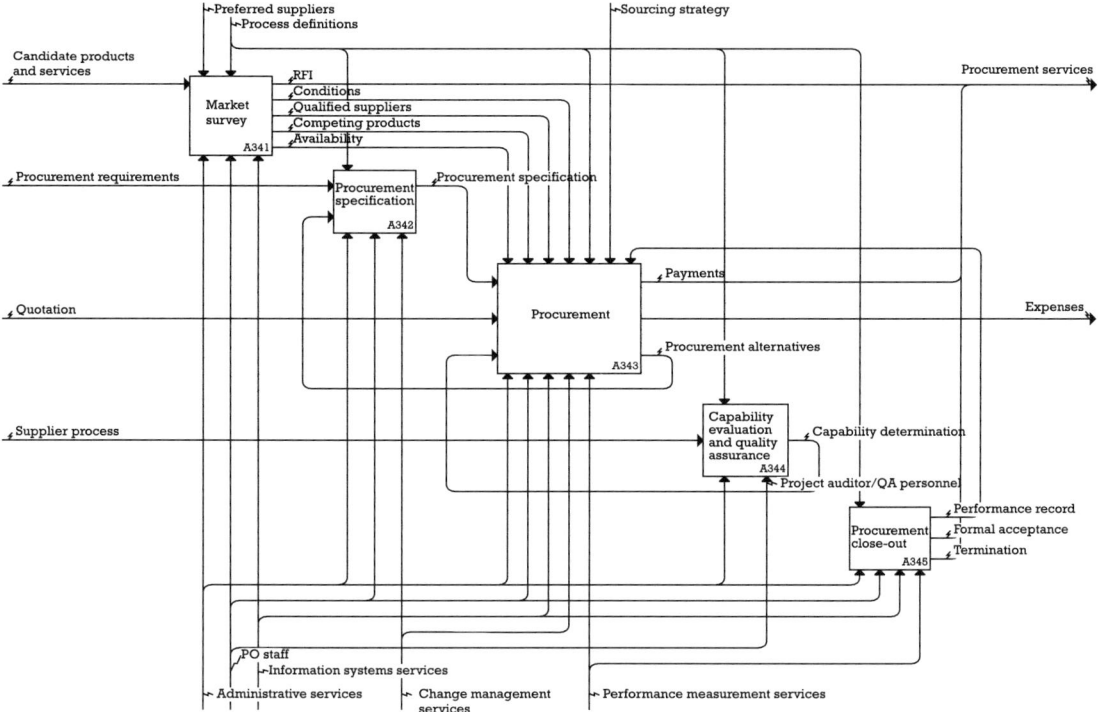

Figure 3.21   Procurement management process.

The QA process (see Figure 3.22) independently and objectively does the following:

▸ Evaluates the quality of work products and ensure consistency with specifications;

▸ Verifies that the work is performed according to the applicable process descriptions and standards;

▸ Provides feedback to project staff and managers on the results of QA activities;

▸ Follows up on noncompliances and ensures that all issues are addressed.

3.3.3.6   Administrative support
This process concerns the administration of the internal PO work. Examples of this are requisition of personnel, travel arrangements, budget preparation, and filing and communication.

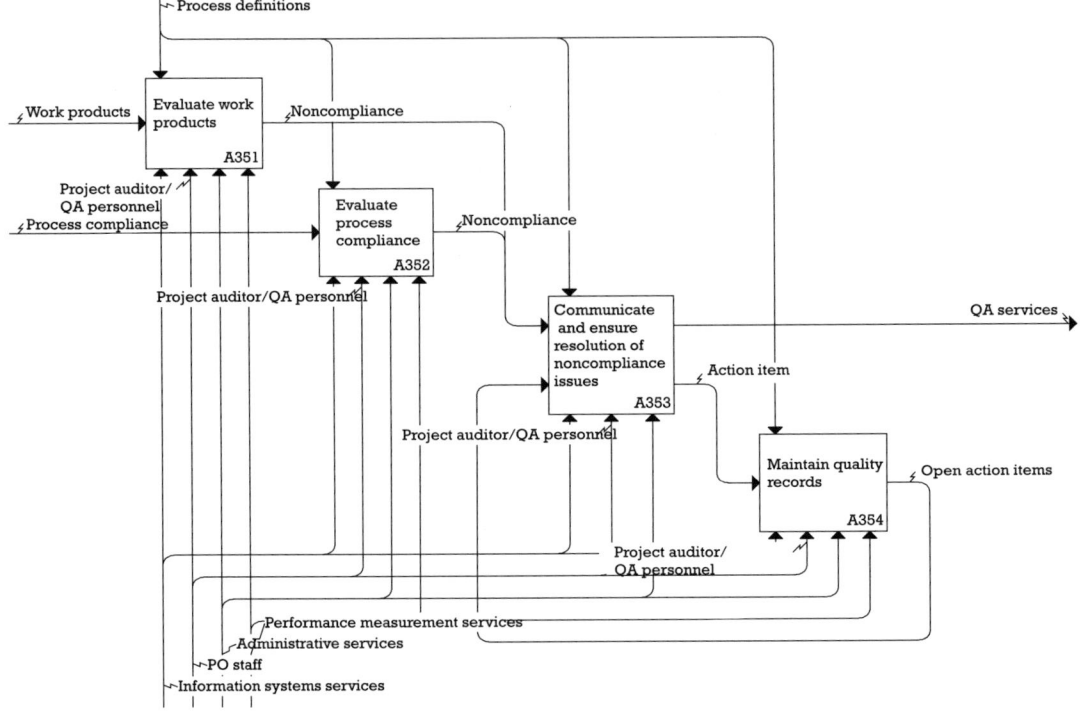

Figure 3.22 Quality assurance process.

3.3.3.7 Project accounting

Project accounting is the process of analyzing, recording, and reporting on all of the financial events originated in a project (see Figure 3.23).

The project accounting process consists of the following activities:

- *Validation:* Validation is not the same as approval. Approval refers to an authorization to spend given by the project manager or other responsible party, while validation is an action performed by the project controller or his delegate to verify that an expenditure conforms to organizational policies.

- *Transaction analysis:* This is the process of deciding which account or accounts should be debited or credited and in what amounts. This is a critical activity for companies benefiting from tax credits, industrial benefits, or any other government incentive program.

- *Burden calculations:* If applicable, a supplement called burden will be added to the base costs for invoicing purposes. Burden costs are

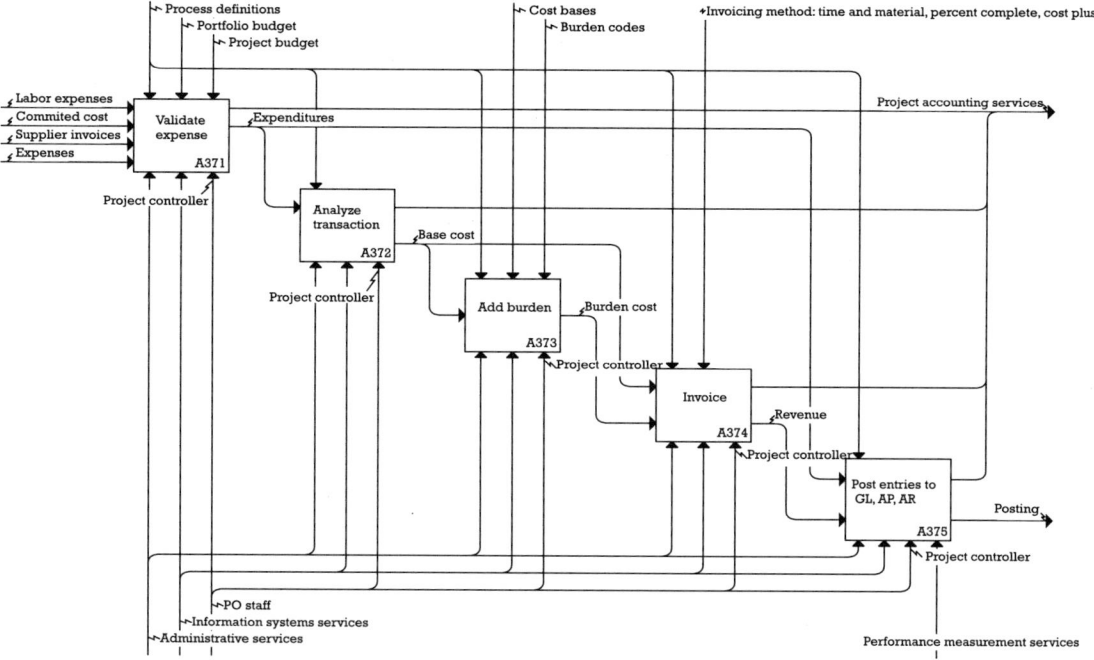

Figure 3.23 Project accounting process.

calculated by multiplying a burden rate, which depends on the type of expenditure (e.g., labor, material), by the base cost.

‣ *Invoicing:* This is the task of generating invoices billed to the project sponsor. The invoicing procedure will depend on the type of contract and the modality of payment agreed upon between the sponsor and the performing organization.

‣ *Posting:* Posting is the process of recording changes in the ledger accounts exactly as specified in the journal entries.

An important but often neglected aspect of project accounting is the definition of account codes useful not only for financial reporting purposes but for managerial reporting as well. For example, time reporting data could be used to determine when a product platform is reaching the end of its useful life by comparing the relative cost of extending its capabilities over successive generations of new products [4]. Chapters 4 and 7 will address this point in greater detail.

3.3.3.8 Human-resources management

The PO is responsible for identifying, acquiring, and developing project management and project support personnel. In order to perform this function, the PO must prepare job descriptions and training programs, and work together with human resources in the establishment of appropriate career paths and rewards mechanisms (see Figure 3.24).

The PO develops the competence of its personnel through job rotation, formal training, self-development, and mentoring and coaching programs.

3.4 PO roles

The exact composition of the PO in terms of the number of personnel, their responsibility assignments, their expertise, and whether they each have single roles or wear several hats depends on the number of projects in the project portfolio, the number of projects in execution at a given time, the projects' size, and the type of PO implemented. Responsibility assignments, however, should not be arbitrary; accountability must go hand in hand with authority and involvement in the decision process. Typical roles that have evolved through the practice of project management are presented below.

3.4.1 PO manager

The PO manager is responsible for running the PO and for the management of the project portfolio. Typical tasks include the following:

- Preparation and maintenance of the organization's master and resource plans;
- Continuous evaluation of project performance to (1) allow the forecasting of future resource needs and (2) highlight areas of deviation where management action is required;
- Recruitment and evaluation of permanent and temporary PO staff;
- Participation in the project's planning sessions;
- Prioritization of efforts and resolution of issues within area of responsibility;
- Preparation of budget, business cases, scenario analysis, contract reviews, and risk-management strategies within area of responsibility;
- Introduction of new technologies and best practices for project management;

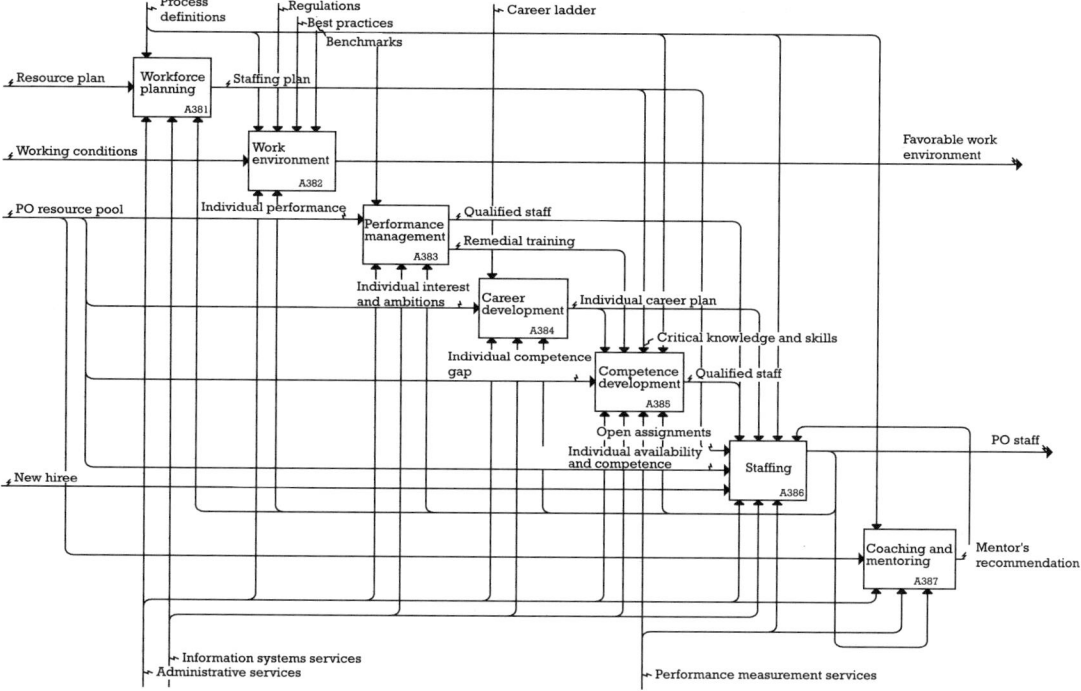

Figure 3.24   Human-resources management process.

- Participation in project steering groups;
- Mentoring of potential project managers;
- Coaching of PO members in the application of the organization's defined processes, methods, and guidelines, and in the use of the organization's tools;
- Facilitation of team meetings;
- Facilitation of sales support;
- Auditing of projects for compliance with guidelines.

The position of PO manager is a very important one, one that due to its characteristics could be used as a training ground for those being groomed for senior management. Besides the technical competencies and the experience necessary to perform effectively in this role, the PO manager must possess business acumen, a network of contacts, the ability to take the initiative when required, an understanding of the points of view of all project stakeholders, and a system-thinking attitude.

3.4.2 Project controller

The project controller is responsible for all project accounting and cost control within the PO. Typically, the project controller will have two reporting lines: one to the PO manager, the other to the organization's controller. More specifically, the project controller provides financial and accounting guidance to the PO and the project managers, and ensures the integrity of the projects' budgets by controlling scope changes, fiscal changes, and overhead allocations, and by flagging significant project overruns and underruns. Typical responsibilities include the following:

- Challenging all inputs to assure their validity and appropriateness;
- Authorizing funds disbursements;
- Establishing procedures for financial reporting;
- Preparing financial reports;
- Providing assistance and expertise related to the organization's financial system;
- Verifying that all expenditures are properly recorded;
- Assisting the project manager in developing the WBS structure to identify the tasks or project elements to be controlled;
- Establishing account numbers for the projects;
- Assisting project managers in the preparation of the project's budgets;
- Identifying and reporting current and future deviations from budgets or other financial problems;
- Assisting the project auditor in the conduct of project audits;
- Conducting follow-ups on contract payments.

3.4.3 Project auditor and quality assurance personnel

The project auditor and quality assurance personnel are responsible for verifying the state of the project based on objective evidence, performing QA tasks, and assessing third-party quality systems. Responsibilities include the following:

- Conducting interviews;
- Analyzing project deliverables;
- Analyzing project data;
- Preparing reports;
- Participating in tollgate decisions;

- Defining opportunities for improvement;
- Conducting root-cause analyses;
- Writing and maintaining the projects' quality plan;
- Developing, adapting, and tailoring development processes;
- Coaching members of the team in the application of the project's processes, methods, and guidelines;
- Facilitating team meetings;
- Promoting process adherence;
- Auditing products for compliance with guidelines;
- Writing action items concerning risks and nonconformances with the prescribed guidelines;
- Collecting project metrics;
- Reporting project metrics.

3.4.4 Project manager

The project manager plans and executes the project on behalf of the project sponsor. To do this, the project manager must coordinate and integrate activities across multiple functional lines. Typical responsibilities include the following:

- Performing key planning work and giving adequate direction to those performing detailed planning;
- Reviewing contracts and proposals;
- Assuring that all goals, plans, and schedules are consistent;
- Establishing and maintaining effective control of the project work and expenses;
- Issuing work guidance;
- Leading the team;
- Promoting a healthy working environment;
- Interfacing with the project sponsor;
- Interfacing with the customer;
- Interfacing with third parties (suppliers and subcontractors);
- Monitoring results to assure that specifications and contract conditions are being met by all parties;
- Controlling changes in the scope of work;

- Participating in risk/opportunity studies;
- Participating in tradeoff studies;
- Authorizing project payments/expenditures;
- Approving project reports.

3.4.5 Project coordinators

The project coordinator assists the project manager in the administration of the project. This position will usually exist only in medium to large projects where the administrative load would distract the project manager from his primary role, or where the organization uses an apprenticeship approach to develop project management competencies. Typical responsibilities include the following:

- Preparing and maintaining the project schedule;
- Preparing and maintaining all the project's correspondence;
- Preparing and maintaining the project's library;
- Preparing and releasing, on approval of the project manager, work authorization documents;
- Maintaining the project ledgers, verification of invoices and their correct holdback, invoice coding, and allocation;
- Obtaining periodic progress reports from all responsible managers;
- Recording the minutes of the project review meetings;
- Following up on action items.

3.4.6 Configuration management personnel

Configuration management personnel are responsible for documenting, monitoring, evaluating, controlling, approving, and communicating all changes made to project charters, the requirements dependency matrix, and any other information shared by more than one individual or organization. Typical responsibilities include the following:

- Organizing and facilitating configuration control board meetings;
- Developing, adapting, and tailoring the project's change management processes;
- Conducting configurations audits;
- Entering and maintaining metadata for configuration items;
- Receiving engineering change proposals.

3.5   Relationships among the PO, the line functions, the project sponsors, and other project stakeholders

Whatever the preferred distribution of responsibility among the PO and other project stakeholders, it is important that none of the tasks falls through the cracks and that everybody understands what is expected of him or her in order to minimize conflicts. An excellent vehicle to achieve this is the responsibility matrix [5], which provides, in a compact form, an unequivocal definition of the authority and responsibility of all the project's stakeholders: senior managers, sponsor, project manager, line managers, PO managers, technical and PO support staff, and so on (see Table 3.1). Different responsibilities allocations would lead to different types of PO. The one shown here corresponds to a managerial type of PO.

3.6   Summary

Chapter 3 introduced the processes necessary to coordinate and support project work and assigned responsibility for them to a new line function, the PO. In practice this framework, like any other framework, must be tailored to the needs and culture of the organization in which it is going to be deployed; this can be done through the use of a responsibility matrix in which the key decisions that must be made through the life of a project are listed and responsibility for them assigned to the various stakeholders.

References

[1]   Archibald, R. D., *Managing High-Technology Programs and Projects,* 2nd ed., New York: Wiley, 1992.

[2]   Goldratt, E. M., and J. Cox, *The Goal: A Process of Ongoing Improvement,* 2nd rev. ed., Great Barrington, MA: North River Press, 1992.

[3]   Capability Maturity Model Integration for Systems Engineering and Software Engineering (CMMISM), Version 1.1, Software Engineering Institute, 2001.

[4]   Meyer, M. H., and A. P. Lehnerd, *The Power of Product Platforms: Building Value and Cost Leadership,* New York: Free Press, 1997.

[5]   Reinertsen, D. G., *Managing the Design Factory: A Product Developer's Toolkit,* New York: Free Press, 1997.

Table 3.1   Responsibility Matrix

	Senior Managers	Project Sponsor	Line Managers	Project Office Manager	Project Manager	Project Office Specialists	Technical Disciplines
Project Formulation				O			
Set project goals		AE			I	I	I
Set project requirements		A			I	I	E
Prepare project schedule		A			E	I	I
Prepare project budget		A			E	I	I
Determine required quality		AE			I	I	I
Determine revenue dependencies		AE			I	I	I
Determine technical dependencies		I			I	I	AE
Determine solution approach		A			A	I	AE
Project Startup				O			
Appoints project manager		A		A			
Modify project requirements	A	AWL		AWL	AWL	I	E
Modify project budget	A	AWL		AWL	AWLE	I	I
Modify project schedule	A	I		AWL	E	I	I
Modify manning plan		I	A	AWL	AWLE	I	I
Select team members			AE	I	I	I	
Select engineering tools			A		I	I	E
Select development methods			A		I	I	E
Make/buy decisions		A			AWL	I	AWLE

Table 3.1 (continued)

	Senior Managers	Project Sponsor	Line Managers	Project Office Manager	Project Manager	Project Office Specialists	Technical Disciplines
Project Execution				O			
Interface with sponsor/customer					E		
Interface with line managers					E		
Modify project requirements	A	AWL		AWL	AWL	I	E
Modify project budget	A	AWL		AWL	AWLE	I	I
Modify project schedule	A	I		AWL	E	I	I
Modify manning plan		I	A	AWL	AWLE	I	I
Remove team member			A		A		
Authorize the use of overtime	A			AWL	AWL		
Authorize travel					A		
Authorize purchases					A		
Approve payments					A		
Project Closure				O			
Hand-over deliverables					A		
Evaluate team member performance			A		E		
Approve lessons learned		I			A	I	I
Grant rewards		I	A		A		I
Conduct Tollgate Review				O			
Evaluate business reality		E			I	I	I
Evaluate project progress		E		I	I		I
Evaluate resource usage		E	I	I	I		I
Approve tollgate		A	I	I	I		I

Table 3.1 (continued)

	Senior Managers	Project Sponsor	Line Managers	Project Office Manager	Project Manager	Project Office Specialists	Technical Disciplines
Project Portfolio Planning				O			
Cancel project	A	I	I	I	I		I
Accepts new projects	A	I	I	I			
Decide growth strategy (hire, outsource, hold, downsize)	A	I	E	I			
Prioritize projects	A	I	I	I			
Resolve escalated issues	AE	I	I	I	I		I
Project Oversight				O			
Orders spot check				A	I	E	I
Approves project diagnostic				AE	I	I	I
Portfolio Control				O			
Authorizes the use of more resources			A	AWL	I		I
Authorizes a schedule extension	A	I	I	AWL	I		I
Authorize the use of reserve funds				A	I		I
Orders a project audit		I		A	I	E	I
Procurement Management				O			
Select sourcing strategy	A	I	I		I	E	I
Select contractors	A	I	I		I	AWLE	I
Select vendors	A	I	I		I	AWLE	I
Negotiate	A	I			I	AWLE	I

Table 3.1 (continued)

	Senior Managers	Project Sponsor	Line Managers	Project Office Manager	Project Manager	Project Office Specialists	Technical Disciplines
Human Resources Management				O			
Recruit PO staff	A			AWL		E	
Evaluate performance of PO staff	A			AWL		E	
Promote PO staff	A			AWL		E	
Terminate PO staff	A			AWL		E	
Project Audit				O			
Process and Information Systems Management				O			
Measurement Process				O			
Change Management				O			
Quality Assurance				O			
Administrative Support				O			
Project Accounting				O			

Legend:

O   Owns process, is responsible for its execution.

A   Approves, is accountable for. More than one A in a row means that it must be agreement.

AWL Approves within limits. If the magnitude of the decision is outside limits, it is referred to A.

I   Input, provides information.

E   Executes, does the actual work.

CHAPTER

4

Processes

Common processes provide the foundation upon which the PO operates. A multiproject environment requires the establishment of common processes for a variety of reasons: First and foremost, they provide the common language indispensable to communication across projects and disciplines and they facilitate the training and integration of new personnel. Second, processes capture the collective knowledge developed through the experience and insight of the PO staff. Third, commonality is essential for the effective use of forecasting models, tools, and databases.

Processes and tools are intertwined. Processes to a large extent determine the choice of tools, but in order to take advantage of a powerful tool, processes that do not provide a definitive benefit must be changed. Both processes and tools are embodiments of the collective knowledge of the organization, and as such their value as a competitive advantage must not be underestimated. As the organization learns, processes, practices, and tools must be changed to reflect new understandings and insights.

In the previous chapter we introduced the main PO processes and interfaces, and the roles necessary to execute them. This chapter will discuss the goals that these processes must achieve and how they can be described.[1] The next chapter will deal with the tools required.

---

1. This chapter will discuss the control arrow labeled "process definition" in the Chapter 3 diagrams. Chapter 5 will address the mechanism arrow labeled "information system services."

4.1 PO process definitions

In the previous chapter, we recommended that the PO owns those processes that clearly fall under its area of responsibility and those that enable it to stay in the loop; that is, those processes by means of which it is decided which projects will be initiated, what changes must be made to the project scope, and how time and expenses will be reported. Its involvement will prevent the PO from being bypassed in these important decisions. Some of these processes might involve adaptations to the project work derived from more general corporate processes, such as purchasing or budgeting.

Process descriptions ought to be written at different levels of detail and using different formalisms. For example, processes like the project portfolio management process, which need to express the flexibility required at the business level, are better served by a description of the policies to be considered in selecting and prioritizing projects than by a prescriptive, step-by-step decision procedure. The opposite is true when it comes, for example, to the CM process. The level of detail present in any process description should be limited to what is needed to achieve the organizational goal of commonality, while preserving the flexibility needed to run projects effectively and efficiently. Overly detailed corporate processes are a barrier to learning and innovation. Table 4.1 presents a process architecture based on these ideas.

Table 4.1 Process Architecture

Process Description	Addresses	Invokes
Project life-cycle management	Project formulation	Estimation
	Project startup	Project auditing
	Project execution	Budgeting
	Project closure	Risk and opportunity management
	Project reviews	
	Tollgate reviews	Quality assurance
Project portfolio management	Project portfolio planning	Budgeting
	Project oversight	Risk and opportunity management
	Project portfolio control	
Estimation	Production of rough and refined estimates	
Budgeting	Itself	
Requirements management	Change management	
Risk and opportunity management	Itself	
Project audit	Itself	

Table 4.1   (continued)

Process Description	Addresses	Invokes
Quality assurance	Evaluation of work products and process compliance	Project audit
	Communication and resolution of noncompliance issues	
	Maintenance of quality records	
Procurement management	Market survey	
	Procurement specification	
	Procurement	
	Capability evaluation and quality assurance	
	Procurement close-out	
Project accounting	Validation of expenses	
	Transaction analyses	
	Invoicing	
	Posting of entries to GL, AP, AR	
Measurement process	Planning of measurement	
	Performing of measurement	
	Production of performance statistics	
Configurations management	Change management	
Human-resources management	Workforce planning	
	Work environment	
	Performance management	
	Career development	
	Staffing	
	Coaching and mentoring	
Administration	Itself	
Process management	Identification of process constraints	
	Minimization of constraints	
	Deployment of process improvements	

4.1.1 Project life-cycle management

The project life-cycle management process defines how projects are formulated, planned, executed, and closed. Specifically, this process defines the following:

- How projects are to be organized and controlled;
- Standardized work breakdown structures and activities;

▶ Which project elements should be reviewed before work is authorized to proceed.

The goals of this process are as follows:

▶ To establish a common vocabulary known to all stakeholders, which enables the use of common tools and reporting routines and the sharing of knowledge across the organization;

▶ To define the interfaces among the various parties involved in a project.

In order to exercise business control over the project, its life cycle will be divided into phases (see Table 4.2), with well-defined checkpoints or tollgates at which the decision to continue or stop the work is formally reviewed. Decisions about the continuation, termination, or change of direction concerning the project work are made by the project's steering committee (see Figure 4.1), made up of senior management, the project sponsor, the PO manager, and the project manager.

When it comes to defining standardized phases and work breakdown structures (WBS) for the purpose of establishing common reporting routines, a balance must be struck between the need for commonality and the

Table 4.2 Project-Management Model

Phase	Formulation	Startup	Execution	Closure
Purpose	To understand the scope of the project and produce order-of-magnitude estimates necessary for evaluating its feasibility	To produce the detailed plans necessary for project execution	To execute the project as planned with respect to time, cost, quality, and scope	To hand over the results, dismantle the project organization, compile a record of all experiences, and bring to a close all outstanding matters
Inputs	Customer or sales requirements Business case	Project charter Historical data Applicable standards	Project specification Contracts Applicable standards	Project deliverables Team feedback Customer feedback
Activities	Establish scope of work Define work approach Produce order-of-magnitude estimates Evaluate risks	Confirm scope of work Produce budgeting estimates Produce detailed plans Evaluate risks	Conduct kickoff meeting Authorize work Monitor progress and expenditures Update plans Communicate project status	Hand over results Document unresolved issues Participate in postmortem reviews Dismantle team Thank those involved
Outputs	Project charter	Project specification	Project deliverables	Lessons learned Updated records

Table 4.2   (continued)

Phase	Formulation	Startup	Execution	Closure
Tollgate			Depending on the duration of the project, there might be more than one tollgate during execution	
Business	The business case remains current	The business case remains current	The business case remains current	All handover has been performed
	The expected project results are in accordance with company strategy	The expected project results are in accordance with company strategy	The expected project results are in accordance with company strategy	All accounts are closed
	The expected project results justify its cost	The expected project results justify its cost	The expected project results justify its cost	
Progress	The selected approach is feasible	The selected approach is feasible	The selected approach will produce a system that meets expectations	The results meet expectations
	Scope of work	Requirements stability	Progress is being made according to the plan	Final report is written
	Commitment	Risk identification	Requirements stability	
			Risk identification	
Use of resources	There are enough resources to execute the project	There are enough resources to execute the project	Staff and equipment availability	
			Resources are consumed according to the plan	
			Estimates to complete are within limits	

uniqueness of the project work. As shown in Figure 4.2, the push for stan-
dardization should not exceed the level of visibility needed by the PO or any
other project stakeholder, leaving to the project team the freedom to further
adapt their activities, within the prescribed categories, according to their
own needs and idiosyncrasies.

Project reviews are an important component of the project life-cycle
process. A project review should not be confused with a project audit. A
project review is a planned risk-reduction activity; a project audit, on the
other hand, is an activity imposed on the project when the circumstances,
usually bad, mandate. The goals of the review are as follows:

- To mitigate risk by finding problems before moving on to the next
  project phase;

- To generate buy-in among the project's stakeholders;

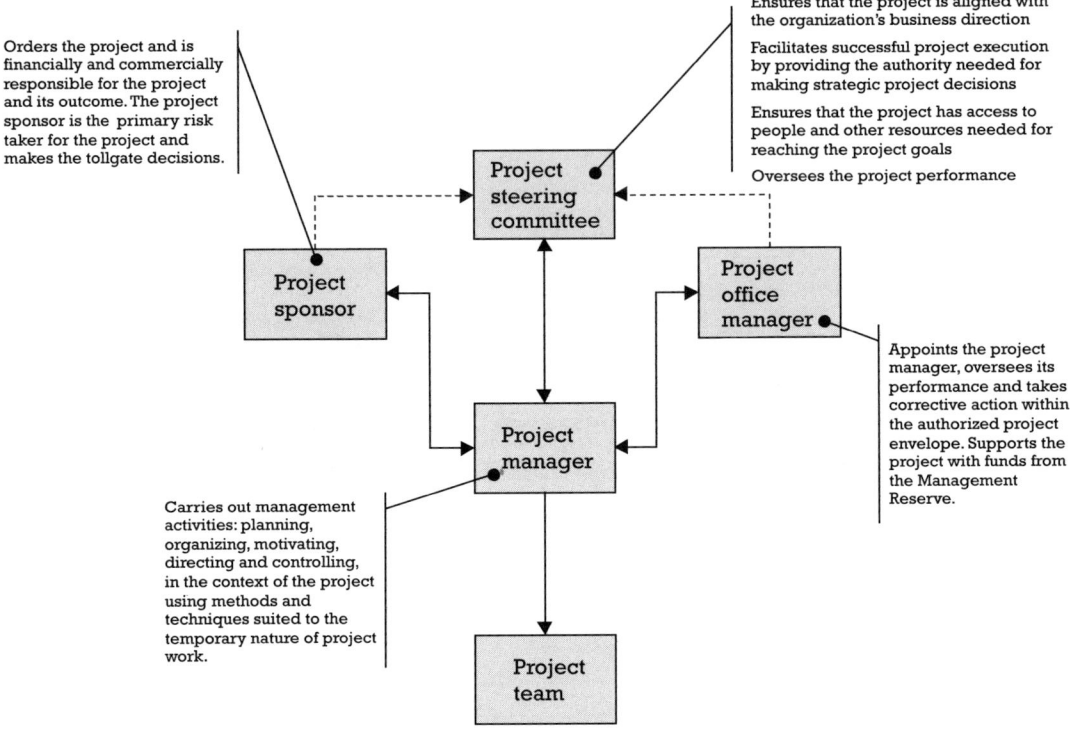

Figure 4.1   Roles in project life-cycle management.

> To identify opportunities that might have been missed in the past.

In order to be effective, project reviews should be based on in-process and final work products, and not on materials generated especially for the review. A good practice is to use a questionnaire or checklist to guide the inquiry, similar to the one shown in Figure 4.3, and to follow up with the following queries [2]:

> How do you know?

> What does that mean?

> Can you show me?

During the review, any issues that could not be resolved within the project should be brought to the attention of the sponsor and the PO manager, and agreed-upon mitigation strategies for near- and long-term risks should be identified.

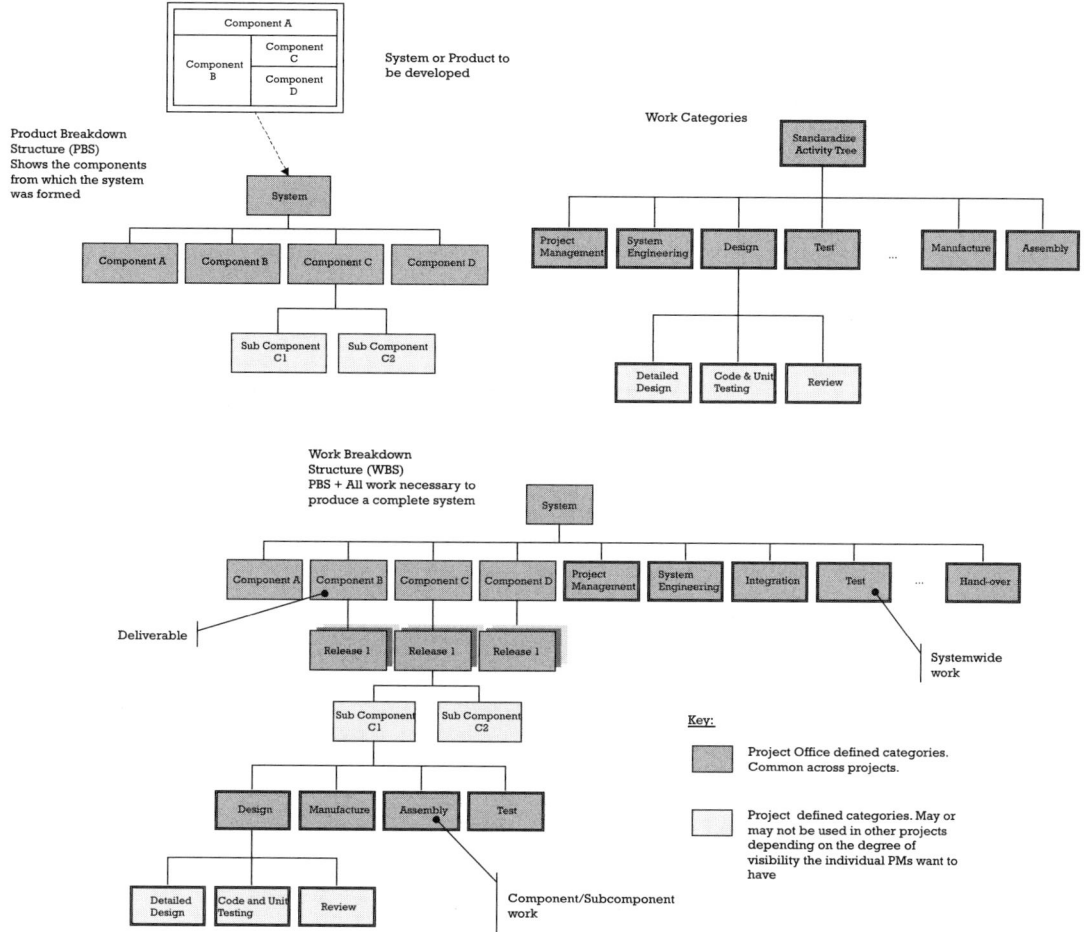

Figure 4.2   Standardized WBS. (*After:* [1].)

4.1.2   Project portfolio management

The purpose of project portfolio management is to advance the goals of the organization by making decisions that take into consideration all, rather than individual, projects that the organization intends to pursue. Principal considerations include the following:

▸ Which combination of projects will maximize benefits?

▸ Which will minimize risk?

▸ Which will most closely align with the company's strategic goals?

Requirements	Management	Work Climate

Stability [Are requirements changing even as the product is being produced?]
• Are the requirements stable?
• Are the external interfaces changing?
(If no) What is the effect on the system?
• Quality
• Functionality
• Schedule
• Integration
• Design
• Testing
Completeness [Are requirements missing or incompletely specified?]
• Are there any TBDs in the specifications?
• Are there requirements you know should be in the specification but aren't?
(If yes) Will you be able to get these requirements into the system?
• Does the customer have unwritten requirements/expectations?
(If yes) Is there a way to capture these requirements?
	Planning [Is the planning timely, technical leads included, contingency planning done?]
	  • Is the program managed according to the plan?
	(If yes) Do people routinely get pulled away to fight fires?
	  • Is re-planning done when disruptions occur?
	  • Are there contingency plans for known risks?
	(if yes) How do you determine when to activate the contingencies?
	  • Are long-term issues being adequately addressed?
	Project Organization [Are the roles and reporting relationships clear?]
	  • Is the program organization effective?
	  • Do people understand their own and others' roles in the program?
	  • Do people know who has authority for what?
		Quality Attitude [Is there a lack of orientation toward quality work?]
		  • Are all staff levels oriented toward quality procedures?
		  • Does schedule get in the way of quality?
		Cooperation [Is there a lack of team spirit? Does conflict resolution require management intervention?]
		  • Do people work cooperatively across functional boundaries?
		  • Do people work effectively toward common goals?
		  • Is management intervention sometimes required to get people working together?
		Communication [Is there poor awareness of goals, poor communication of technical information among peers and managers?]
		  • Is there good communication among the members of the program?
		  • Managers
		  • Technical leaders
		  • Developers

Figure 4.3 Project review lines of inquiry concerning requirements, management, and work climate. (*After:* [3].)

> • Which projects need to be started and when?
> • Which ones need to be terminated?

Specifically, this process addresses the following:

> • How project requests are received and evaluated;
> • The criteria to be applied in deciding whether or not to incorporate or remove a project from the portfolio;
> • The preparation of the master plan, the strategic resource plan, the financial forecast, and the requirements dependency matrix;
> • How projects are started and terminated;
> • The leeway or margin of maneuver that the PO has in addressing deviations in the project portfolio.

Rather than defining painfully detailed procedures unlikely to be followed by senior management and project sponsors in their decision-making process, this process must clearly state the business and project information required to support well-thought-out portfolio decisions. It must also define the latitude, in terms of the projects' scheduling and effort windows, that the PO manager has in making decisions without approval from senior management and project sponsors. If there is little latitude and the PO manager is forced to go back to management and sponsors every time a decision needs to be made, the PO loses much of the reason for its existence.

Since the consequences of many of the decisions made at the portfolio level will take months or maybe years to be felt, it is critical that the process be supported by decision aids such as system dynamics [4] to evaluate the long-term consequences of decisions, the analytical hierarchical process [5] to help in the selection of projects based on multiple criteria, and Monte Carlo simulation techniques to deal with uncertainty [6]. These techniques and others for portfolio planning will be explained at length in Chapter 6.

4.1.3   Estimation

The estimation process defines the method or methods used to size and estimate the time and effort required to do a job. The estimates produced become the basis for project scheduling and resource allocation. Estimates are revisited throughout the project's life cycle, typically at major milestones or when significant changes to requirements or project constraints impose a revision of the original plans.

The estimation process involves the following:

▸ Sizing the job;

▸ Finding the cost drivers;

▸ Forecasting the time and effort required;

▸ Conducting sensitivity analysis.

The goals of this process are as follows:

▸ To produce accurate estimates of the resources and time required to complete a given work element;

▸ To have documented estimates that can be reviewed and scrutinized by others.

A thorough discussion of the estimating process would require a separate book. However, for our purposes we will focus on two key observations. The first is that estimates are invariably required before all the necessary information is available; the second is that no matter how much effort is put into an estimate, there will always be variances as a result of the number of variables that can influence the effort and time required to perform a task. Project work is after all a stochastic rather than a deterministic process. The corollary to these two assertions is that there will be multiple estimates along the life cycle of a project, each with different degrees of accuracy as a function of the information available (see Table 4.3), and that every estimate should be qualified with a probability statement (see Figure 4.4) that reflects the likeliness of achieving it. We will return to this in Chapter 6.

4.1.4 Budgeting

The budget prepared by the PO is derived from the level of project activity that the organization is planning to sustain in the budgeting period. The budget will identify the source and destination of the funds administered by the PO. The structure associated with the budget process in shown in Figure 4.5.

Similar to the financial forecast described in Chapter 3, a budget is a plan expressed in monetary terms covering a specified period of time, usually 1 year. However, unlike the financial forecast, which is merely a prediction of what people think might happen used for decision-making purposes, the budget carries the implicit commitment of the PO manager and the project managers to take positive steps toward making sure that the budgeted events actually happen.

To keep the master plan and the budget consistent, the PO will maintain a rolling budget. This means that every quarter, a new budget will be prepared by dropping the amounts for the quarter just completed, reviewing the amounts for the succeeding three quarters, and then adding the amounts corresponding to the fourth succeeding quarter. The goals of the budgeting process are as follows:

▸ To give the PO manager a say in major project decisions.

▸ To delegate authority to project managers and PO manager, enabling them to spend budgeted funds without seeking approval.

Table 4.3 Different Types of Estimates

When?	Type of Estimate	Level of Detail (WBS level of decomposition)	Estimating Method	Accuracy (%)
Feasibility, prestudy	Order of magnitude	1	Parametric, analogy, paired comparisons	±35
Planning	Budget	2, 3	Successive principle, analogy, parametric, Paired comparisons	±15
Execution	Definitive	4, 5, 6	Successive principle, engineering buildup	±5

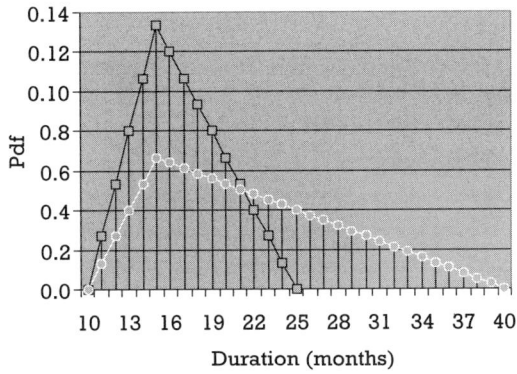

(a) Probability density function

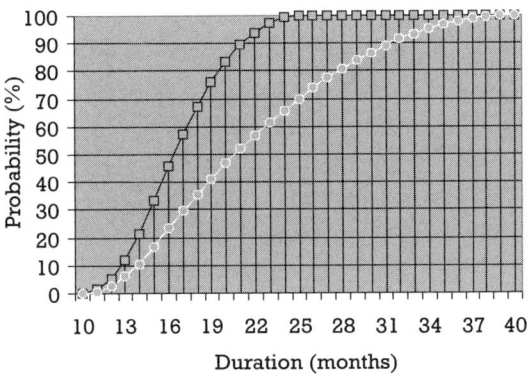

(b) Cumulative distribution

Figure 4.4 Two projects with identical best-case (10 months) and most likely (15 months) duration scenarios, but different worst-case duration scenarios (25 versus 40 months), have very different on-time probabilities (in the first project, the probability of finishing within or before 15 months is 34%, and in the second, it is 18%).

‣ To establish a yardstick against which the actual performance of the project managers and the PO manager will be measured.

Project managers are responsible for producing and administering the project budgets, while the PO manager prepares and administers the total budget and the management reserve. The rationale for putting the management reserve under the control of the PO manager instead of under the control of each project manager is that by spreading the risk across the portfolio, individual projects are afforded a given level of protection at a lower cost [7, 8]. This will be explained in detail in Chapter 6.

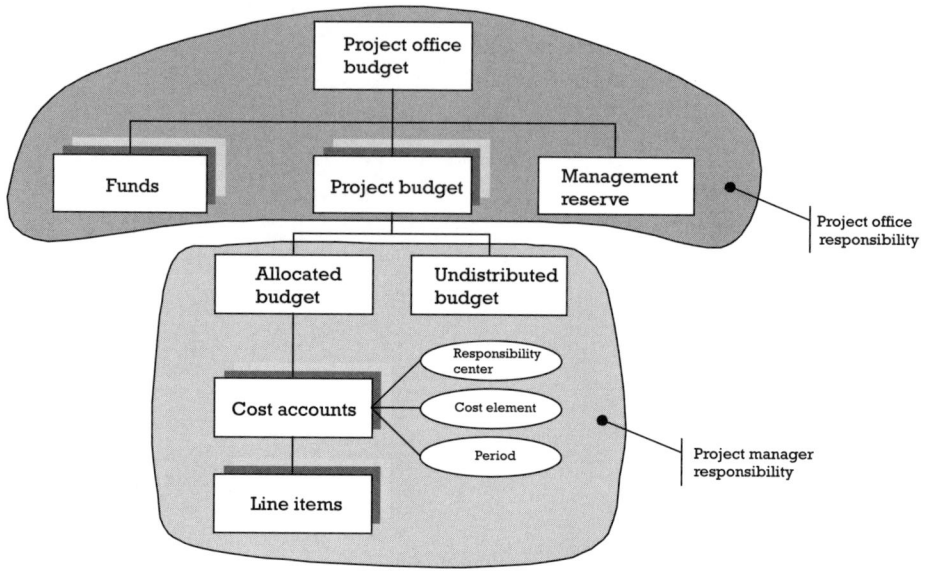

Figure 4.5   Budget structure.

4.1.5   Requirements management

Requirements management involves establishing and maintaining an agreement between the project sponsor and the supplier organization. The agreement includes business, technical, and performance information about the work products that will be delivered as a result of the project work. The agreement forms the basis for estimating, planning, performing, and tracking project' activities. As new requirements are added or existing requirements deleted or modified, the project budget and its schedule are revised.

The requirements management process addresses the following:

 ▸ Definition, documentation, and verification of the scope of work;
 ▸ Requirements metadata (ancillary information that is an aid to understanding, prioritizing, and controlling the requirements);
 ▸ Traceability between business cases, scope of work, and work products;
 ▸ Handling of changes in the scope of the work.

The goals of this process are as follows:

- To assure that there is a known baseline for engineering and management use;

- To assist in the identification of items, whether business cases or work products, likely to be affected by a change.

In addition to the traditional practice of maintaining traceability between requirements and work products, the PO maintains traceability between requirements and between requirements and business cases (see Figure 4.6). The traceability between requirements enables the PO to evaluate the consequences to the rest of the portfolio of dropping or postponing the fulfillment of a requirement. The traceability between requirements and business cases allows the PO to determine which projects are likely to be affected in case the assumptions on which the business cases rest do not hold.

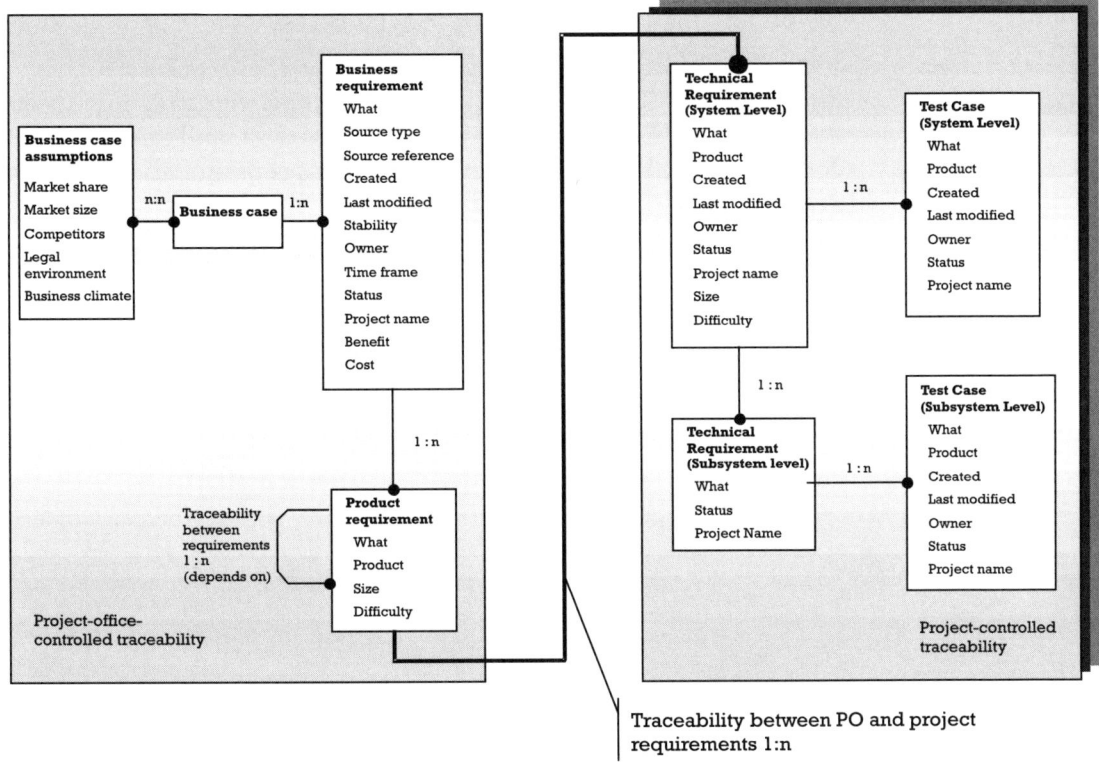

Figure 4.6   Requirements management schema.

4.1.6 Risk and opportunity management

Traditionally, risk management in projects has been concerned with avoiding the risks that might jeopardize success from within the project. The portfolio approach, however, presents new possibilities. Today, the project portfolio approach opens the door to a different interpretation of risk management, an interpretation that, as in finance, reflects the connection between risk and returns. This new interpretation focuses not on avoidance, but on actively managing the risks that must be taken in the pursuit of opportunity and, ultimately, profit.

From the project perspective, the risk-management process establishes how project risks are evaluated, documented, and mitigated (see Figure 4.7). The risk-management process involves the following activities:

▸ *Risk planning:* This is the up-front activity necessary to execute a successful risk-management plan. The planning should assign responsibility for specific risk-management functions and establish risk reporting and documentation requirements. Should the conditions in the project change drastically, it may be necessary to replan the risk-mitigating actions.

▸ *Risk assessment:* This activity includes the identification of critical issues that might have an adverse impact on the project, and an analysis to determine the likelihood of occurrence and its consequences. The risk-assessment method is shown in Figure 4.8.

▸ *Risk handling:* After the project's risks have been identified and assessed, the approach to handling each significant risk must be developed. There are essentially four strategies for handling risk: avoiding it, controlling it, transferring it, or assuming it. For all identified risks, the best strategy should be evaluated by looking at its feasibility, expected effectiveness, cost and schedule implications, and the effect on the system's technical performance.

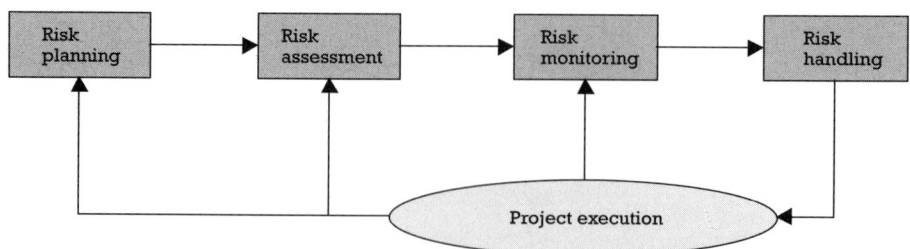

Figure 4.7 Risk-management process.

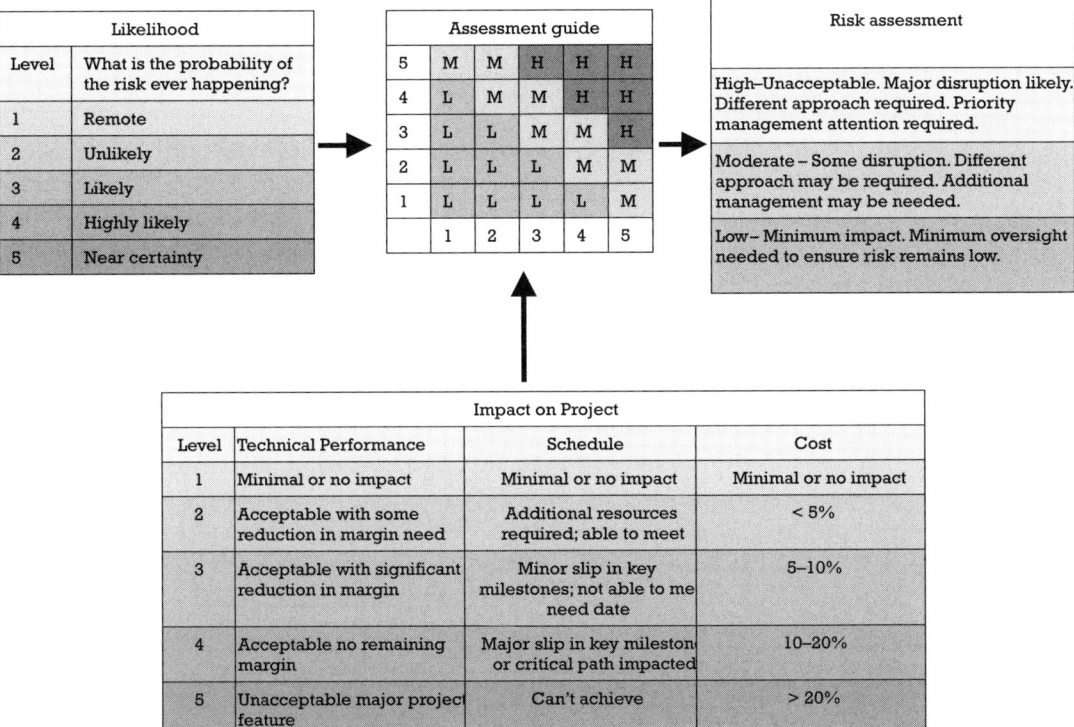

Likelihood	
Level	What is the probability of the risk ever happening?
1	Remote
2	Unlikely
3	Likely
4	Highly likely
5	Near certainty

Assessment guide

5	M	M	H	H	H
4	L	M	M	H	H
3	L	L	M	M	H
2	L	L	L	M	M
1	L	L	L	L	M
	1	2	3	4	5

Risk assessment

High–Unacceptable. Major disruption likely. Different approach required. Priority management attention required.

Moderate – Some disruption. Different approach may be required. Additional management may be needed.

Low – Minimum impact. Minimum oversight needed to ensure risk remains low.

Impact on Project			
Level	Technical Performance	Schedule	Cost
1	Minimal or no impact	Minimal or no impact	Minimal or no impact
2	Acceptable with some reduction in margin need	Additional resources required; able to meet	< 5%
3	Acceptable with significant reduction in margin	Minor slip in key milestones; not able to me need date	5–10%
4	Acceptable no remaining margin	Major slip in key mileston or critical path impacted	10–20%
5	Unacceptable major project feature	Can't achieve	> 20%

Figure 4.8   Risk-assessment method. (*After: Risk Management Guide for DoD Acquisitions*, 4th Edition, Defense Acquisition University Press, Feb. 2001)

> • *Risk monitoring:* This is the systematic tracking and evaluation of the performance of the risk-handling actions. Essentially, it compares predicted results of planned actions with the results actually achieved to determine status and the need for any change in risk-handling strategy.

From a portfolio perspective, the PO must address two types of risks:

1. Private or diversifiable risks (i.e., schedule slip, unreliable technology, staff turnover, misunderstood business);

2. Market or nondiversifiable risks (i.e., economic health, competing standards, war).

The PO addresses private or diversifiable risks by spreading them across the project portfolio (for example, by charging insurance premiums proportional to the risk exposure of the projects). Risk spreading not only reduces the cost of capital, as the organization's immobilized capital is abridged and the need for last-minute funds is minimized, but could also be converted into a competitive advantage by using it in bidding work or by including price incentives in contracts. In order to work, the process must produce fair and consistent assessments of the true project risks. This could be achieved, for example, by applying risk taxonomies such as the ones presented in Tables 4.4 and 4.5, in combination with a quantitative assessment of the probability of occurrence of a given risk and its impact in monetary terms on the project budget.

The tollgate model by which the project is authorized to proceed addresses the market or nondiversifiable risk arising from the uncertainty with respect to the project's payoff. By breaking down the development into phases, the tollgate model provides management with the flexibility to delay committing funds until the uncertainty is resolved at a minimum cost. Similar to a financial option, which management might exercise but is not obliged to if the circumstances are not favorable, at each tollgate management would decide whether to continue, kill, or redirect the project in response to new information not available at the beginning of the project. This greatly enhances the project's value by improving its upside potential while limiting its losses to the cost of the preceding project phases.

Project insurance and real option valuation will be explained in more detail in Chapter 6.

4.1.7 Project audits

The project audit process establishes the steps to be followed and the artifacts—documents and deliverables—to be inspected with the purpose of establishing the true and fair status of a project with the goal of independently assessing the extent to which the original business objectives could be achieved and the cost of recovery.

Despite their procedural similarity, reviews and audits are quite different. Reviews are scheduled activities, established with the purpose of finding inconsistency or missing items before they cause problems. Audits on the other hand are unplanned activities performed when a project has gone awry and it is necessary to conduct a major replanning before deciding on the continuation or termination of the project.

The audit should, at a minimum, consider the following:

Table 4.4   Risk as a Function of Technology Maturity

	Industry		NASA's Technology Readiness Level (TRL)
	Software	Pharmaceutical	
Highest risk	Papers published, university originated tools	Synthesis and extraction	TRL1—Basic principles observed and reported
	Books and commercial tools available (Release 1)	Biological screening and pharmacological testing	TRL2—Technology concept and/or application formulated
	Books and commercial tools available (Release 2+)	Pharmaceutical dosage formulation and stability testing	TRL3—Analytical and experimental critical function and/or characteristic proof-of-concept
	At least one commercial product exists that uses the technology		
	Industry and/or institutional standards start to be developed	Toxicology and safety testing	TRL4—Component and/or breadboard validation in laboratory environment
	Training becomes widely available		
	Predominant design established	Investigational new drug (IND) application	TRL5—Component and/or breadboard validation in relevant environment
		Phase I clinical evaluation	TRL6—System/subsystem model or prototype demonstration in a relevant environment
	Industry and/or institutional standards are accepted	Phase II clinical evaluation	TRL7—System prototype demonstration in a space environment
	Implementation feasibility is high	Phase III clinical evaluation	TRL 8—Actual system completed and "flight qualified" through test and demonstration (ground or space)
	Technology is scaleable, replicable, and extensible		
		Process development for manufacturing and quality control	
		Bioavailability studies	
		New drug application (NDA)	TRL 9—Actual system "flight proven" through successful mission
Lowest risk	Technology vendors start to add "proprietary features" to the standard implementations (Release 3+)	Postapproval research	

Table 4.5   Software Risk Taxonomy (*After:* [3])

Product Engineering	Development Environment	Program Constraints
Requirements Stablility Completeness Clarity Validity Feasibility Precedent Scale	*Development process* Formality Suitability Process Control Familiarity Product Control	*Resources* Schedule Staff Budget Facilities

Design
Functionality
Difficulty
Interfaces
Performance
Testability
Hardware
  constraints
Nondevelopmental
  software
	Development system
	Capacity
	Suitability
	Usability
	Familiarity
	Reliability
	System Support
	Deliverability
		Contract
		Type of contract
		Restrictions
		Dependencies

Code and unit test Feasibility Testing Coding and Implementation	*Management process* Planning Project organization Management experience Program interfaces	*Program interfaces* Customer Associate contractors Subcontractors Prime contractor Corporate management Vendors Politics
Integration and test Environment Product System	*Management methods* Monitoring Personnel management Quality assurance Configuration management	
Engineering specialties Maintainability Reliability Safety Security Human factors Specifications	*Work environment* Quality attitude Cooperation Communication Morale	

‣ Business situation;

‣ Critical-path status;

‣ Milestone hit rate;

‣ Deliverables status;

‣ Requirements stability;

‣ Critical technologies readiness;

> Cost to complete;

> Actual resources versus planned resources;

> High-probability, high-impact risk events;

> General disposition of the team;

> Sponsor's commitment.

The audit process should include interviews with the people doing the work and the project sponsor. The audit process should also provide recommendations with respect to the composition of the audit team and the circumstances under which it should be initiated.

4.1.8   Quality assurance

The purpose of quality assurance (QA) is to provide management with oversight of the project process and the products being built. In subcontracting and outsourcing situations, the QA function will also be responsible for the qualification of subcontractors and external providers.

QA involves inspecting products and activities performed to verify that they comply with the applicable procedures and standards. Unlike a review, where the main question is "Are we doing the right things?" a quality inspection focuses on the question "Are we doing things right?"

Compliance issues must first be addressed within the project. If, for whatever reason, this is not possible the issue should be referred to the appropriate management level for resolution. Key elements of this process are the escalation procedure, which must be documented in order to prevent ill-feelings between QA and the project staff, and the existence of an independent reporting channel to the PO manager and senior management, as these are two of the few instruments the QA function has to leverage its authority. The following is a nonexclusive list of activities to be performed by the QA group.

> Ensure that the work is conducted in accordance with the standards and procedures established by the project and as described in the project charter; raise and follow up on noncompliances;

> Ensure that life-cycle documents and the requirements dependency matrix are prepared and kept current and consistent;

> Verify that relevant life-cycle documents are updated and based on approved requirements change;

> ‣ Identify defects, verify resolution for previously identified defects, and ensure change control integrity.

> ‣ Selectively review and audit the content of system design and other project documents.

> ‣ Ensure that action items resulting from reviews of the software requirements analysis are resolved in accordance with these standards and procedures.

As quality is built into the product rather than added at the time of project completion, the QA work is an ongoing process that continues throughout the project life cycle.

4.1.9 Procurement management

Many of today's projects and products are so complex that few organizations have all the necessary product and process knowledge required to completely design and manufacture them in house. As a result, most companies are dependent on others for crucial elements of their corporate offerings. Typically, however, companies have some choice as to which providers they become dependent upon and for what sorts of products, skills, and competencies [9].

Procurement management is the process that deals with the procurement of goods and services from third parties. The major issues addressed by this process are make/buy decisions, vendor solicitation and selection, and contract negotiation and awards. Specifically this process addresses the following:

> ‣ Market surveillance;

> ‣ Market investigation;

> ‣ Solicitation;

> ‣ Contract tracking and oversight;

> ‣ Evaluation;

> ‣ Transition to support.

While certain process activities such as market surveillance and market investigation will be entrusted to technical specialists from the line functions, others like solicitation, because of its legal and financial implications, will be assigned to dedicated personnel with specialized knowledge. The knowledge required to perform this function ranges from the selection of the type of contract (e.g., fixed price, cost plus incentive) under which the work will be developed, to the negotiation of data rights, penalties, payment schedules, fee structures, scheduling of the deliveries, and warranties.

The goals of the external provisioning process are to do the following:

‣ Select dependable suppliers and partners;

‣ Minimize life-cycle costs;

‣ Develop lasting relationship with suppliers and partners.

At the PO level, it is likely that the procurement process will be executed in the context of a higher-level acquisitions process that provides strategic direction concerning what should be outsourced, who the preferred partners are, and what conditions must be negotiated.

4.1.10   Project accounting

The aim of a specialized project accounting process is to eliminate the need for cumbersome subaccounts in the general ledger while giving the project personnel full access to critical cost and budget analysis information. Among the mismatches between corporate and project accounting, we can mention the annual accounting cycle versus the project life cycle, and the need to classify the expenses along dimensions other than responsibility center or job orders.

This is especially important in the case of effort data, frequently the largest cost element in R&D projects. Besides the obvious need to track expenditures for accounting purposes, effort data could be used for a variety of other purposes: to evaluate project progress, to estimate future work, to detect problems, to measure productivity, and to make end-of-life decisions. It is for this reason that it is essential that the employees working in projects report time in a consistent and timely manner and at a level of granularity that matches the planning detail [10].

For each reported hour it should be possible to identify the following:

‣ *Work product:* This dimension serves not only progress and cost reporting purposes, but also has an important use as a normalizing factor in conjunction with other measurements such as deliverables size to calculate productivity and defect density numbers.

‣ *Activity:* This refers to the type of activity (i.e., system design, coding, testing) to which the hours apply. The main purpose of this classification is to provide data for process improvement activities and estimation of future work.

> *Organization:* This corresponds to the cost center to which the employee performing the work belongs and is mainly used for responsibility accounting.

> *Period:* This attribute identifies the time interval, usually a week, in which the hours were worked. This categorization allows the calculation of a number of time-phased metrics such as earned value, but is also important when used jointly with production metrics such as number of line of code or installations performed to calculate the rate of progress.

> *Type of labor:* This category is used to differentiate between direct or billable hours and indirect hours such as those devoted to process improvement, training, and other support activities. This division is important not only for the different accounting implications but as a measure of the slack or white space that exist in the organization. As we saw in Chapter 2, an organization with no or very little slack is likely to enter into a fire-fighting mode in response to minor disturbances.

> *Type of hours (i.e., regular or overtime):* Independently of whether it is paid or not, overtime must be recorded. This is necessary in order to prevent other metrics from being biased because of inaccurate reporting of the actual time it takes to perform a certain activity or complete a product. Overtime is also a leading indicator of morale problems in the project team.

> *Main purpose of the work (i.e., development or rework):* This helps to calculate the cost of quality and can be used for process-improvement purposes.

It goes without saying that employees should not be required to enter explanatory data according to each of these categories, but rather that the codes used to report time should contain sufficient information to identify them.

4.1.11 Measurement process

Measurements constitute the fact base on which portfolio management and process improvement rely. Without measurements, there is usually no way to know the status of a project along its many dimensions: cost, schedule, product performance, supportability, and quality. Thus, it is difficult to know what actions to take, what decisions to make, or how to correct unexpected outcomes. Measurement is also becoming increasingly important in two-party business agreements, where it provides a basis for specification, management, and acceptance criteria [11].

The use of measurements in portfolio management will be addressed at length in Chapter 7.

4.1.12   Configuration management

CM is a pervasive discipline that touches on every aspect of the project work. With respect to the project sponsor, CM deals with changes to the project scope; internally, it deals with the evolution of the project's work products. Simply stated, the purpose of CM is to maintain in a congruent state: plans, contracts, requirements, specifications, and deliverables.

Traditionally CM has involved four interrelated tasks:

1. Identification, consisting of the selection of the work products to be controlled and a labeling schema;

2. Change control, comprising all steps necessary to establish or change a baseline;

3. Status accounting, which involves the recording and reporting of the information concerning the status of the work products as well as the changes requested;

4. Audits, to verify the conformance of the work products to their reported status.

In this proposal we include two new efforts:

1. Version control, to maintain a list of up-to-date work products to be utilized for everyday work;

2. Communications, shared with the requirements management process, to ensure that all interested parties are informed of the disposition and consequence of proposed changes.

As shown in Figure 4.9, there can be up to three CM levels: the PO level, which deals with baselines and changes that have repercussions across the portfolio; the project level, which is responsible for the changes to its own development baseline; and the product level, which kicks in after the handover of the deliverables to the project sponsor. The existence of three CM

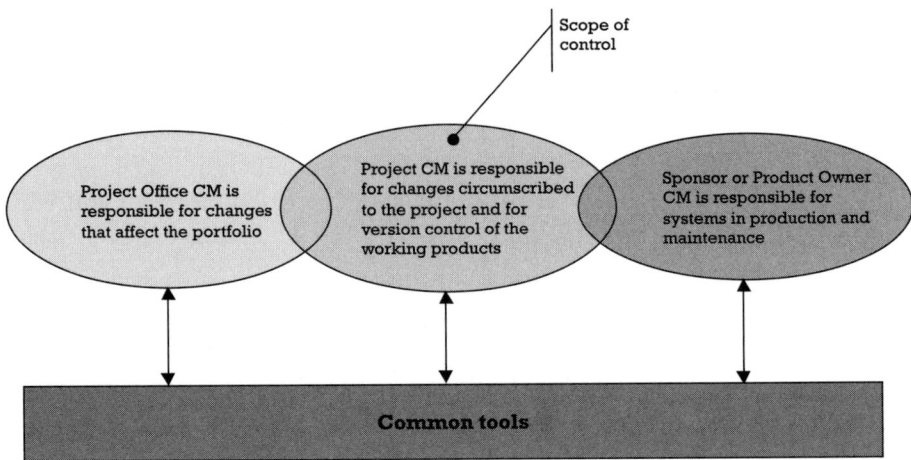

Figure 4.9 CM levels.

levels does not imply in any way the existence of three different systems or three different CM tools. What the three levels do indicate is that there are different groups that will make decisions about different things, and that there are different procedures to respond to each group's needs.

4.1.13 Human-resources management

Usually the human-resources management process will follow organization-wide practices, with the PO left to specify the competencies and skills required and the development of appropriate training plans and career paths. The main purpose of this process with regard to the PO is to continually enhance the ability of the PO workforce to perform their assigned tasks and responsibilities while ensuring that individuals are provided with opportunities that enable them to achieve their career objectives.

Specifically this process establishes the following:

▶ Competency categories under which the knowledge and skills of the project managers are classified;

▶ Work and development activities that contribute to the project manager's development and career growth;

▶ Knowledge, to be acquired via training or self-actualization courses that correspond to each competence category.

Figure 4.10 shows the competence model developed by the Ericsson Project Management Institute, which requires certification by the Project Management Institute or equivalent formal training in the nine project management knowledge areas [12] for individuals wanting to pursue a project management career within Ericsson. This initial certification is later complemented with Ericsson's own training and through coaching and mentoring programs.

Mentoring involves sharing experience and knowledge with others with less seniority through informal methods. A mentorship is a one-to-one relationship in which the mentor serves as a role model for his or her protégé and provides advice on work-related issues and on career development goals and strategies. Coaching is a similar activity but with emphasis on the group rather than on the individual. Coaching is normally done in the context of a supervisor-employees relationship.

4.1.14 Process management

Process management defines how all the other processes owned by the PO are to be defined, who has the responsibility for approving changes to

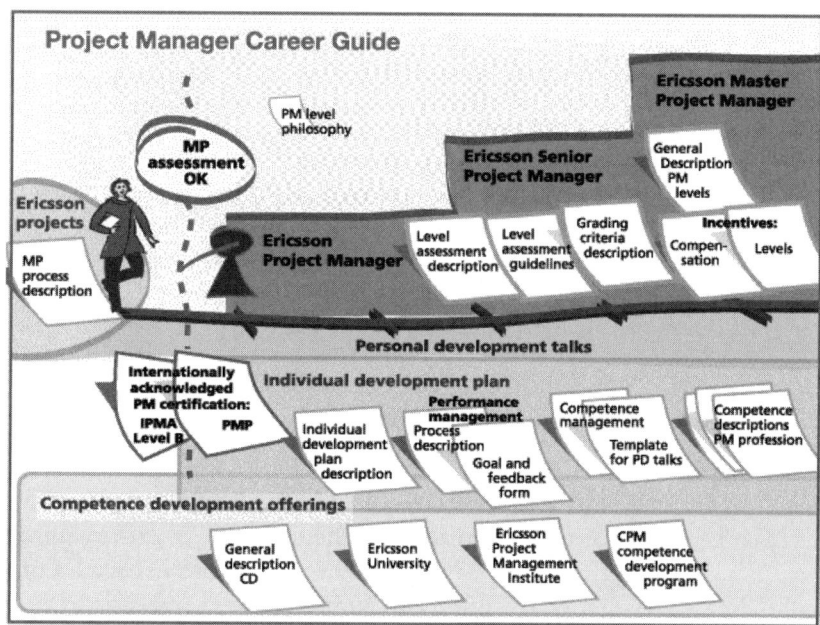

Figure 4.10 Ericsson competence model. (*Source:* [13].)

existing ones, and more important, commits the organization to a practice of continuous improvements. The process management process usually follows the practices and formats established by the organization's management system, especially if the organization is certified under an international standard such as ISO 9000.

There is a tendency in the project management community to associate process work with bureaucracy and delays, as indicated by the not infrequently heard phrase, "Do you want me to follow the process or get the job done?" To counter this perception, processes should be written with the goal of providing a shared understanding about how work is organized, who does what, and how activities within the process interface. Large binders prescribing step-by-step behavior gather dust on the shelves of many organizations. Process descriptions should not become textbooks; they are not intended to replace training or thinking. The process descriptions should provide enough information to answer the questions raised in Tables 4.6 and 4.7. In addition, tailoring guidelines on how to adapt the generic processes to the needs of the individual projects must also be provided.

As much as possible, process knowledge should be embedded in the tools used by the project staff so that process enactment becomes a byproduct of their use, not an intrusion into the staff's work. An example of this would be the replacement of a manual archive and distribution-list system by a central repository with e-mail notifications of all relevant changes, such as the addition of a new document or the creation of a new baseline, to a list of subscribers. Other examples would include the creation of pay or time reporting codes directly from the project plans, the preparation of progress reports from data collected as the result of work transactions such as checking-in or checking out a document from a repository, and the triggering of alerts as a result of the breaking of a predefined business rule associated with a project database, or a process description displayed as part of the Help function in a tool.

4.2 Summary

Processes not only provide the foundation on which the PO operates, they represent the collective knowledge and insights developed through the work of the organization. The strength of an organization lies not only in the technical knowledge and resources it possesses, but in its ability to make these work together and improve them over time.

Table 4.6 Process Description Metadata

Attribute	Meaning	Example
Name	What is the name of the process?	Portfolio planning process
Purpose	Why is the process performed?	To further the goals of the organization by taking decisions in light of all, rather than individual, projects
Input	What work products are used?	Project charters, master plan, resource plan
Output	What work products are used?	Updated master plan, updated resource plan
Role	Who (or what) performs the activities?	PO manager calls and facilitates process. Senior management and project sponsors are responsible for the ultimate disposition of the projects
Activity	What is done?	Review progress for existing projects. Review new projects for inclusion in the portfolio. Project prioritization and resource balancing. Plans updates
Entry criteria	When (under what circumstances) can the process begin?	Quarterly or when a major variance forces a portfolio replanning
Exit criteria	When (under what circumstances) can the process be considered complete?	Agreed master and resource plans
Reviews and audits	List of reviews, checks, and audits performed during the process	N/A
Measurements	Description of measurements applicable to the process	Time-to-market Pipeline index
Cost drivers	Factors that drive the expense of an activity or resource	Number of projects
Training	List of required training for the process	N/A
Tools	List of tools that support the process	Expert choice; MS project
Best practices	List of the things that work, list of things that were tried and did not work	Analytical hierarchical process(AHP), risk planning

References

[1]   NASA System Engineering Handbook, SP-610S, 1996.

[2]   *Little Yellow Book of Software Management Questions,* Software Program Managers Network, 1997.

[3]   Taxonomy-Based Risk Identification, CMU/SEI-93-TR-6, Software Engineering Institute.

[4]   Sterman, J., *Business Dynamics: Systems Thinking and Modeling for a Complex World,* Boston: Irwin/McGraw-Hill, 2000.

Table 4.7   Work Product Description Metadata

Attribute	Meaning	Example
Name	What is the name of the work product?	Project charter
Purpose	What is the purpose?	To specify the project outcome in terms of time, costs, quality, and deliverables. It provides a complete description of the project and serves as a foundation for the project work. The project charter ensures that the business agreement between the project sponsor and the execution organization(s) is clearly defined and described.
Identification conventions	Unique identifier used within the company for CM and filing purposes.	N/A
Table of contents	Description of the information to be provided.	1. Business Direction 1.1 Project Goals 1.2 Business Opportunity 1.3 Project Background 1.4 Project Scope 2. Project Outcome 2.1 Deliverables 2.2 Quality Objectives 2.3 Included/Excluded 3. Plans 3.1 Time-Schedule 3.2 Milestone Definitions 3.3 Delivery Plans 3.4 Project Budget 4. Project Organization and Stakeholders 4.1 Project Management Function 4.2 Project Steering Function 4.3 Project Sponsor

[5]   Saaty, T., *Fundamentals of Decision Making and Priority Theory with the Analytic Hierarchy Process,* Pittsburgh: RWS Publications, 1994.

[6]   Grey, S., *Practical Risk Assessment for Project Management*, Chichester, NY: Wiley, 1995.

[7]   Kitchenham, B., and S. Linkman, "Estimates, Uncertainty, and Risk," *IEEE Software,* May 1997.

Table 4.7 (continued)

Attribute	Meaning	Example
		5. Project Execution
		5.1 External Suppliers
		5.2 Connections to Other Projects
		5.3 Reporting and Communication
		5.4 Reviews
		5.5 Support Activities
		5.6 Security
		5.7 Configuration Management
		5.8 Quality Verification Model
		5.9 Nonconforming Products
		5.10 Risk Management
		5.11 Subcontract Management
		5.12 Document Control
		5.13 Quality Records
		6. Risks and Opportunities
		7. Intellectual Property Rights
		8. Project Handover
		9. Other Matters
Producer	What process produces it?	Project formulation
Consumer/s	What process consumes it?	Portfolio planning; project planning
Date	Date on which work product was produced	
Approver	Who has responsibility for the content of the work product?	
Change history	When, why, who, and what was changed since the previous revision of the work product?	
Status	What is the current state of the work product?	

[8]   Garvey, P. R., *Probability Methods for Cost Uncertainty Analysis: A Systems Engineering Perspective,* New York: M. Dekker, 2000.

[9]   Fine, C. H., and D. E. Whitney, "Is the Make-Buy Decision Process a Core Competence?" Cambridge, MA: MIT Center for Technology, Policy, and Industrial Development, Feb. 1996.

[10] Goethert, W., E. Bailey, and M. Busby, *Software Effort and Schedule Measurement: A Framework for Counting Staff-hours and Reporting Schedule Information,* Software Engineering Institute, CMU/SEI-92-TR-21, 1992.

[11] FDIS 15939: Software Engineering—Software Measurement Process, ISO/IEC JTC1/SC7 N2560, 2002-01-10.

[12] *Guide to the Project Management Body of Knowledge, A PMBOK® Guide,* Project Management Institute, 2000.

[13] Ericsson Project Management Institute, 2002.

CHAPTER

5

Contents

Tools

A the number of projects managed by an organization grows larger and the projects become more complicated, the existence of an adequate information system to support operations is essential.

Tools for project management have long focused on the scheduling of tasks, but the PO needs more information than that yielded by task scheduling. An adequate information system for a PO must provide, in addition to the traditional scheduling capabilities, explicit functionality to monitor the status of resources across projects, and what-if capabilities to support portfolio analysis.

Although the functionality they offer might seem similar, not all tools are the same, nor can the PO delegate responsibility to the IT department for deciding which ones should be used. Information systems can provide organizations with a competitive edge, not only because they enable them to do things that could not otherwise be done, but because the tools themselves become knowledge containers. Through customization and the encoding of business practices, tools make the experience and insight of your best contributors operational throughout the organization.

This chapter begins with the presentation of an idealized information system for the project-based organization and concludes with a survey, necessarily incomplete, of commercial tools that approximate the functionality described here.

5.1 Information needs

A PO information system (see Figure 5.1) must satisfy the needs of three groups of users: the PO and senior manager, concerned with projects and competencies, the project managers, who deal with tasks and generic resources, and the line managers, whose concern is with tasks and named resources. But whatever the user group, the same basic questions must be answered [1]:

- What needs to be done?
- How much will it cost?
- When can it be done?
- What resources will be utilized?
- What are the consequences of doing A instead of B?
- Where are we in relation to where we planned to be at this point?
- Is work progressing at an acceptable rate?
- Where are we going to be a month from now?

These simple questions define the basic functionality that any PO information system must provide.

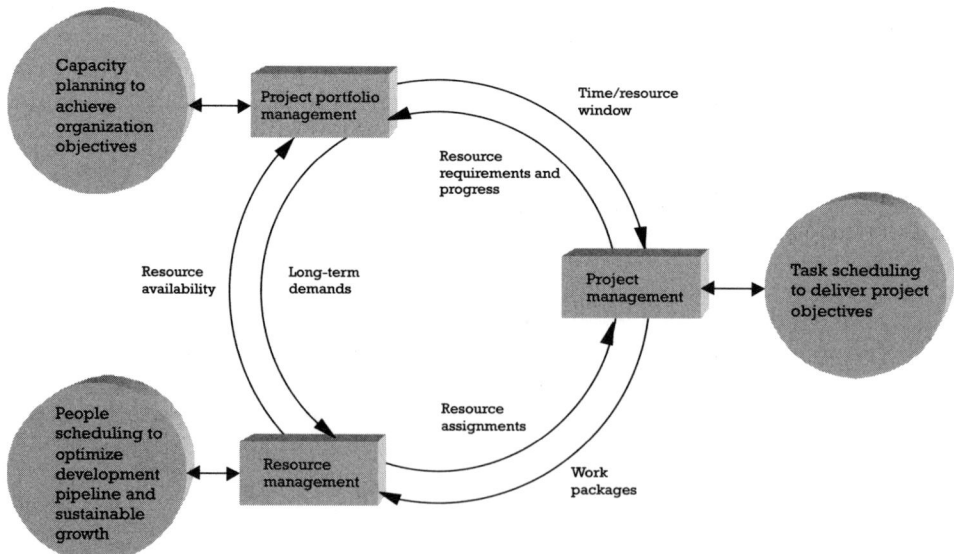

Figure 5.1 Different stakeholders have different objectives but all can be reduced to four basic information needs: What? When? Who? Where are we?. (*After:* [2].)

By *information system* we do not imply a single application package provided by a single vendor or a homegrown monolithic application. The PO
information system could very well be made up of a mixture of commercial
tools and some internal development. What distinguishes a bunch of tools
from a system is that in the latter all the tools have the same understanding
of what the data represents, they share it, and users are not forced to enter it
more than once. If users need to copy and paste between dissimilar applications, or if there exist copies of the same data in slightly different formats to
satisfy the needs of different applications, the organization does not have a
system.

5.2   Characteristics of a PO information system

From a technical perspective, a PO information system should have the following characteristics:

- *A single point of entry:* Data needs to be entered only once, whereon it is
  made available to all the components that require it.

- *Data integration:* The various tools making up the system have access to
  all data. Data integration requires that the data being shared have the
  same meaning, representation, and units across tools. For example, if
  two tools share a piece of data called "task duration," it is crucial that
  both tools use the same calendar and that both agree with respect to
  whether the duration is expressed in working or calendar days.

- *Control integration:* This refers to the possibility of one tool being able to
  invoke operations—for example, sending mail, raising a flag, or updating a data element in response to a change—on a second tool.

- *Presentation integration:* This addresses the need for a common look and
  feel across modules and tools. This characteristic is key for the acceptance of tools. There are few things as annoying as having to learn and
  unlearn a new set of rules every time one switches applications.

- *Analytic and aggregation capabilities:* Since different levels of the organization require information with different levels of detail, it is crucial that
  the information system be capable of filtering and aggregating data
  across multiple dimensions.

- *Openness:* This supports the capability of adding functionality, such as a
  risk module or a critical-chain planning extension, to the core tools via
  the acquisition of specialty add-ons or macro programming, or to add
  new data fields to the database to address the particular information

needs of each project. Taking this approach allows the PO to cater to different needs while attaining tool commonality.

- *Interactivity:* Very few managers will rely solely on the output of an optimization algorithm in making a decision. The output of resource-balancing algorithms, for example, rarely produces a plan that an experienced project manager will consider acceptable. Experienced managers employ recognition-based reasoning in making decisions [3], so it is important that the tools support this cognitive strategy by providing some results, preferably in graphical form, then asking the user for some additional input or guidance, which is then used in generating new output. The process is repeated until the user is satisfied.

- *Exception reporting:* An adequate information system will integrate multiple sources of information to provide a composite picture of what is going on within projects or with respect to the portfolio, and will report exceptions to customizable or user-defined business rules via e-mail based on a subscriber paradigm similar to that used by many Web-based news and trading services.

- *Security and views:* Different people (i.e., project managers, portfolio managers, and resource managers) need information arranged and aggregated in different ways. The information system should support the way people work by providing different views consistent with the role of each user, rather than by forcing them to align their thinking with incompatible views or go through complicated workarounds to gain access to the information they need. Similarly, access privileges (i.e., who has the right to read, add, change, or delete information) must be centrally controlled and based on the roles different users play.

5.3 Functionality of a PO information system

For the purposes of description, in the discussion below, tools are grouped according to their functionality and primary use and explained by way of screen shots from an idealized information system.[1] The purpose of using an idealized, instead of a real, system is to present the reader with a forward view, a vision of what could be, rather than to limit the discussion to the tools available today.

---

1. Some of the tools shown have already been built and are in use at some divisions within Ericsson; others are mockups or composites of existing tools.

5.3.1   Portfolio management

The portfolio management tool group addresses the functionality required to plan and control the project portfolio (see Figures 5.2 through 5.6). It includes the following:

- Master plan preparation;
- Resource plan preparation;
- Financial forecasting;
- Risk-exposure calculations;
- Multicriteria decision support;
- Portfolio control.

At this level, the tool or tools deal with projects and competencies. The workload and financial forecasts are based on templates or pre-established profiles and not on the detailed planning of tasks. Similarly, resource availability is established at the competence level and not for individual resources.

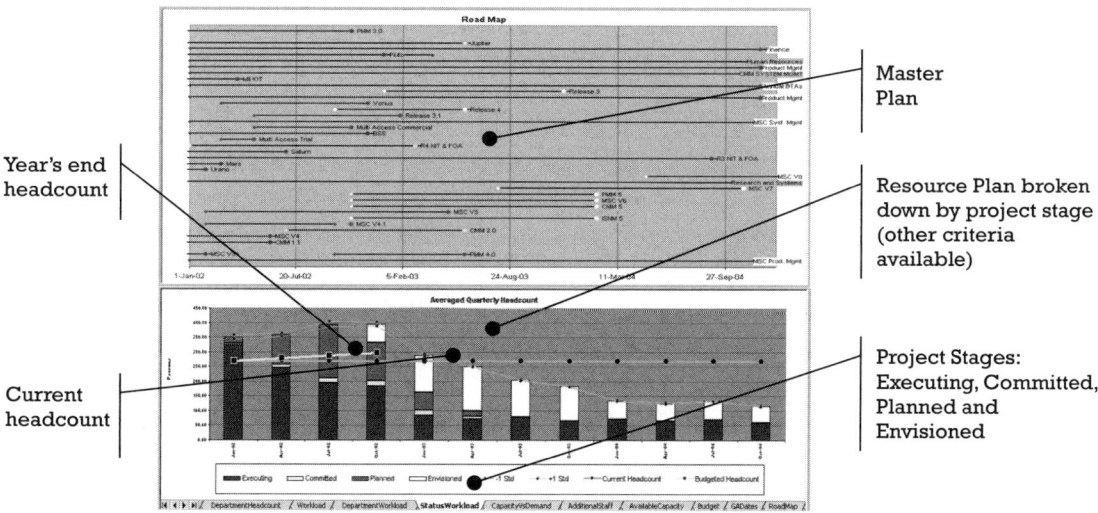

Figure 5.2   Master and resource plans: The workload curve changes in response to the movement of the projects on top.

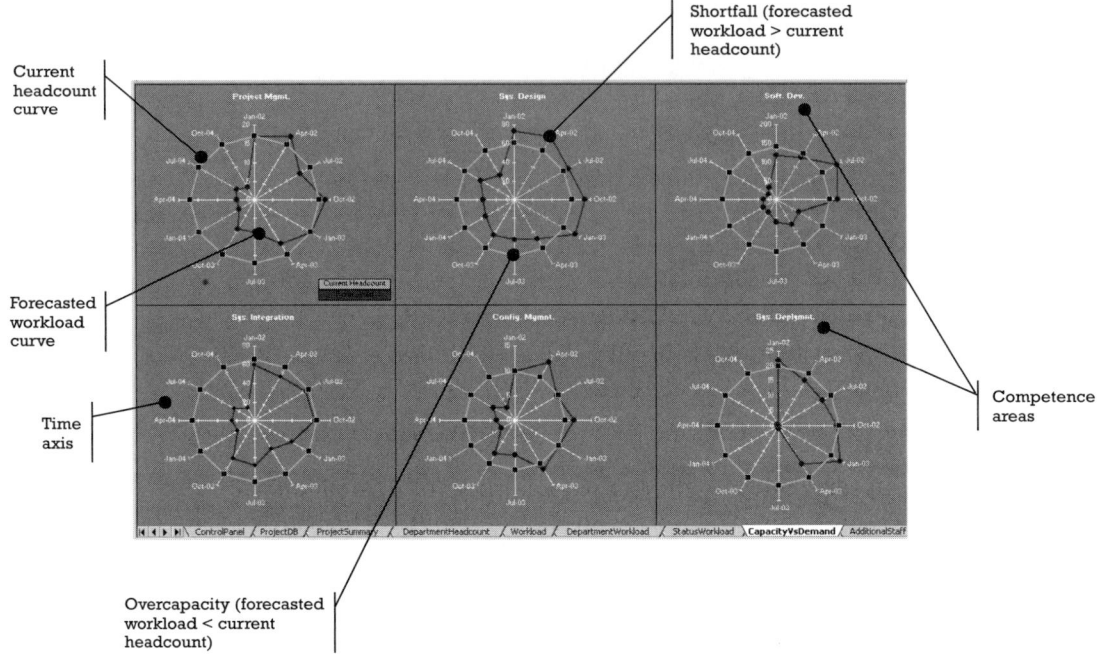

Figure 5.3   Capacity-versus-demand charts.

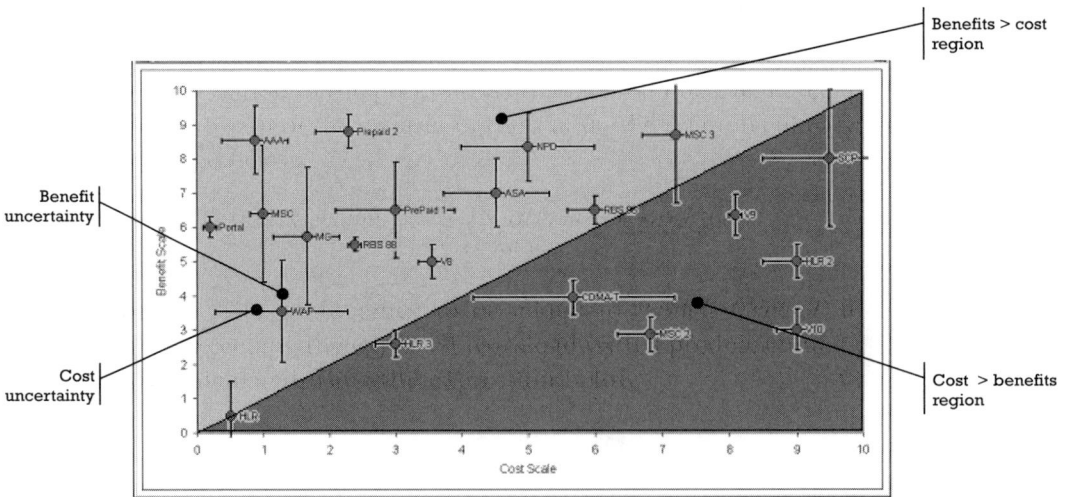

Figure 5.4   Project positioning chart (benefits versus cost).

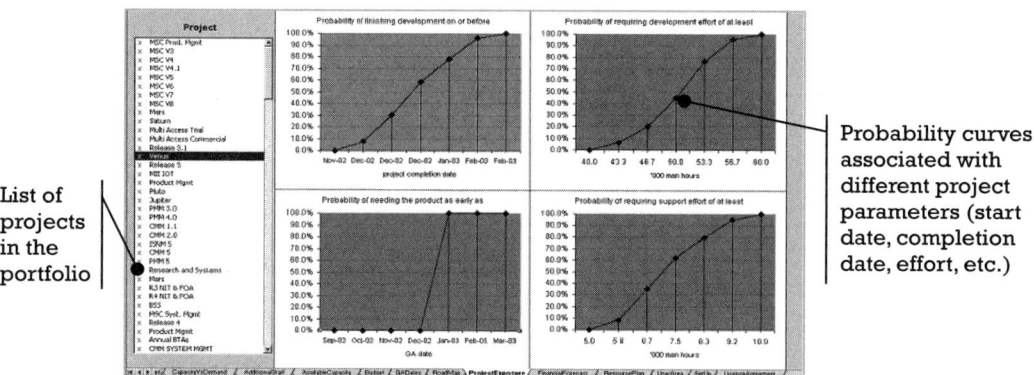

List of projects in the portfolio

Probability curves associated with different project parameters (start date, completion date, effort, etc.)

Figure 5.5 Project risk exposure.

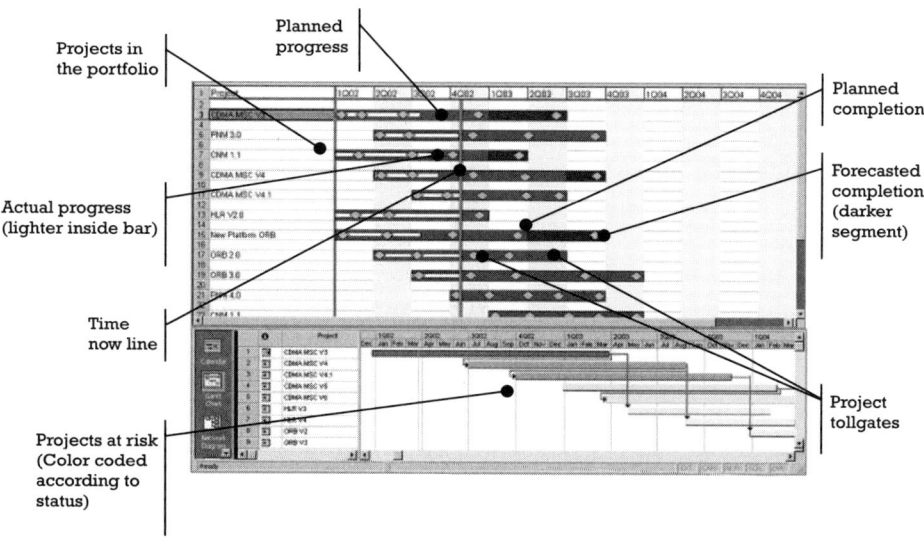

Projects in the portfolio

Planned progress

Planned completion

Actual progress (lighter inside bar)

Forecasted completion (darker segment)

Time now line

Projects at risk (Color coded according to status)

Project tollgates

Figure 5.6 Portfolio tracking.

For the master plan, the resource plan, and the financial forecast, the tool must be able to group or display the data according to criteria such as the type of project (i.e., new platform development, product extension, research) and the degree of project commitment (i.e., whether the project is under execution, approved but not started, or only proposed). This functionality is critical when it comes to prioritizing or rearranging projects to satisfy resource or budget constraints.

Due to the uncertainties associated with long-term planning, it is imperative that the tool or tools used to implement this functionality allow for the probabilistic treatment of the quantities involved. A fundamental capability of these tools is the ability to deal with what-if scenarios and sensibility analyses.

The multicriteria decision support helps the decision makers to rank the projects according to different factors of merit, with the purpose of selecting those that contribute the most to the organizational goals.

For the control part of the portfolio management view, the forecasted completion date of the ongoing projects is contrasted with their planned duration in the master plan, and changes in the resource load and start date of related projects attributed to these differences are flagged. The overall status of individual projects is represented using a traffic-light schema, supplemented with drill-down capabilities to provide visibility into the issues or measurements from which the status is derived.

5.3.2   Task scheduling

This group of tools provides the functionality typically encountered in traditional planning tools augmented with workload and probability curves to help plan under resource or time constraints. The Gantt view (see Figure 5.7) is used to define and sequence work within a project, while the resource-allocation form allows the project manager to allocate or request generic resources (i.e., based on competence and location, rather than specific individuals) to work on specific tasks.

The workload chart, which allows the project manager to quickly visualize the amount of resources required to execute the plan, together with the probability chart, which shows the likelihood of completing a given task or project by a certain date, provides the project manager with the capability to interactively create a resource-constrained plan while making informed choices about the risk exposure associated with committing to a certain completion date.

5.3.3   Resource scheduling

This group provides the functionality required by line managers to efficiently utilize the resources under their supervision.

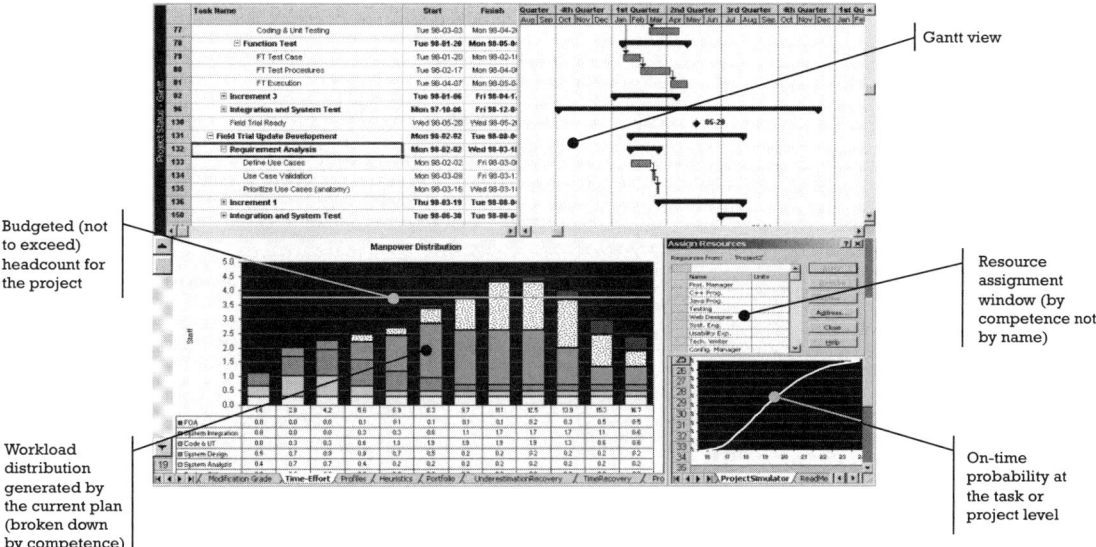

Budgeted (not to exceed) headcount for the project

Workload distribution generated by the current plan (broken down by competence)

Gantt view

Resource assignment window (by competence not by name)

On-time probability at the task or project level

Figure 5.7   Project planning, Gantt view.

In contrast with the resource planning and allocation of generic resources done by project managers, this function deals with the day-to-day assignment of individual resources to specific tasks. For example, the resource manager will look at the requirements for resources generated by the project managers and assign specific individuals to do the work. To perform this task effectively and efficiently, the resource managers need to know the following:

▸ Which tasks require assignment of resources;

▸ The availability of each resource;

▸ The resources at risk of not been released on time;

▸ The task or tasks whose initiation is at risk as a consequence of overallocation or of resources not being released on time;

▸ The aggregated workload for a given period.

The resource-allocation display (see Figure 5.8) allows the line manager to see at a glance: the request for resources, the tasks at risk, and the work that each resource has been allocated to, including indirect activities such as training or vacation.

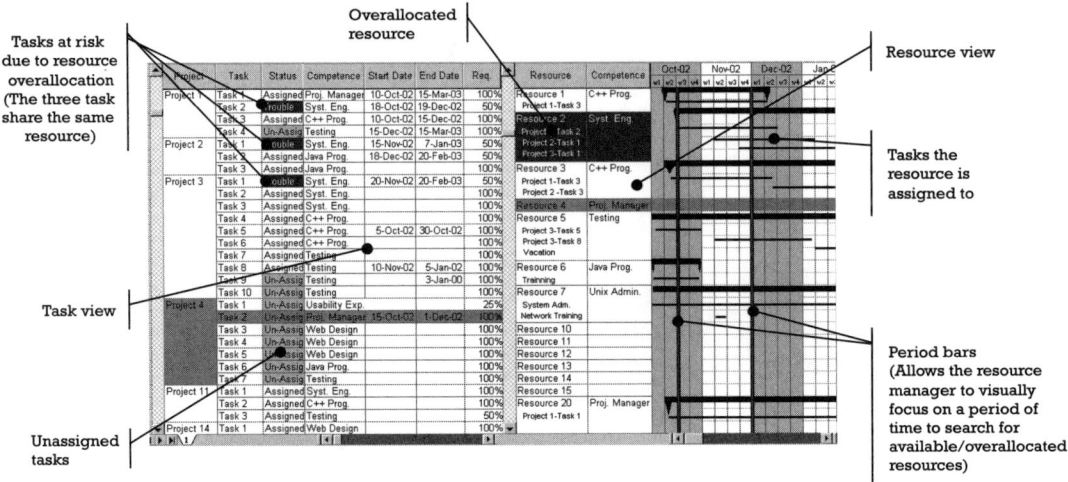

Figure 5.8   Resource allocation display.

5.3.4   Project tracking

Project tracking functionality is based on the control-panel [4] concept in which a central display presents all the information necessary to establish the health of the project and from which it is possible to drill down into different views to get more detailed information about a particular area or issue (see Figure 5.9).

The project tracking system should not be limited to the reporting of past information, but ought to make extensive use of forecasting models to identify the early signs of a delay while there is still time to do something about it [5].

The use of "traffic lights" to bring attention to plan deviations is highly recommended; however, the coloring schema and the meaning of the lights should be used consistently throughout the system to prevent false alarms and misunderstandings. An example of a criterion for setting the red, yellow, and green indicator would be as follows: A red light is used to signal the existence of several indicators outside range or a strong variance in one of them. A yellow light signals mild deviations or isolated readings. A green light signals that the project is progressing according to the plan.

5.3.5   Document management

This functionality provides all project personnel, subject to security restrictions, with a central point for storing and retrieving project documentation.

Indicators showing the information requested by the PM. The side by side display allows the comparison of key variables.

List of recent project events, such as reaching a milestone, releasing a new version of a document, etc. The PM subscribes to the class of events he is interested on.

Controls allow the PM to choose the information and level of detail to be displayed

List of actions which require action. (Color coded according to severity)

List of tasks at risk. (Color coded according to severity)

Gantt view showing the tasks at risk. The completion dates are forecasted using process models.

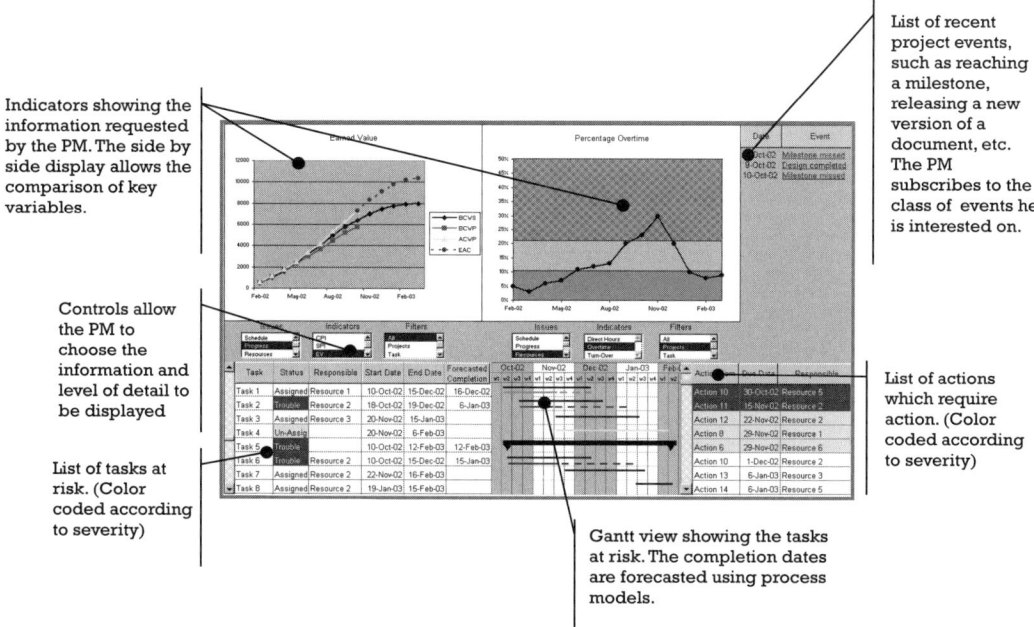

Figure 5.9   Project tracking tool.

Basic capabilities include browsing, classification, and content search. Typical documents stored or accessible through this function include plans, specifications, contact list, minutes of meetings, action items, process descriptions, change requests, correspondence, contracts, and presentations (see Figure 5.10).

5.3.6   Risk management

The risk-management function helps project managers identify, prioritize, and communicate project risks using a structured approach.

This tool provides standard database functions for adding and deleting risks, together with specialized functions for identifying, prioritizing, and retiring project risks. Each risk can have a user-defined risk-mitigation plan and a log of historical events. Additionally, the risk-management tool provides automatic notification and escalation capabilities (see Figure 5.11).

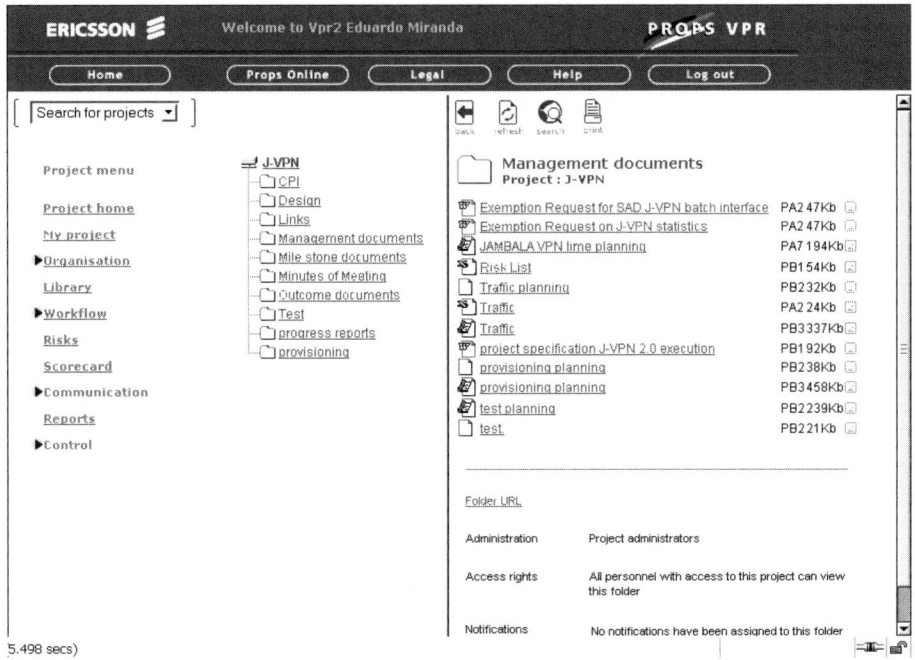

Figure 5.10 Sample document management tool: Ericsson's Virtual Project Room (VPR).

5.3.7 Finances and budgeting

The functionality of this group concerns the preparation of budgets, the tracking of expenses and revenues, the analysis of funding sources and destinations, the rate of expenditures, variances, cash flows, and cost and schedule performance indexes (CPI and SPI) for all project work. See Figures 5.12 and 5.13.

At a minimum, these tools need to support the data structures necessary to roll up the data along project deliverable and organizational lines and provide baselining capabilities so that project performance can be evaluated against what was planned. A good tool would add to this basic functionality the capability of preparing and tracking activity-based budgets [6].

Typically, the tools will be required to deal with the following budget categories: direct labor costs, general and administrative expenses, material, equipment, travel, and management reserves. For large multinational organizations, the ability to deal with multiple currencies is also critical.

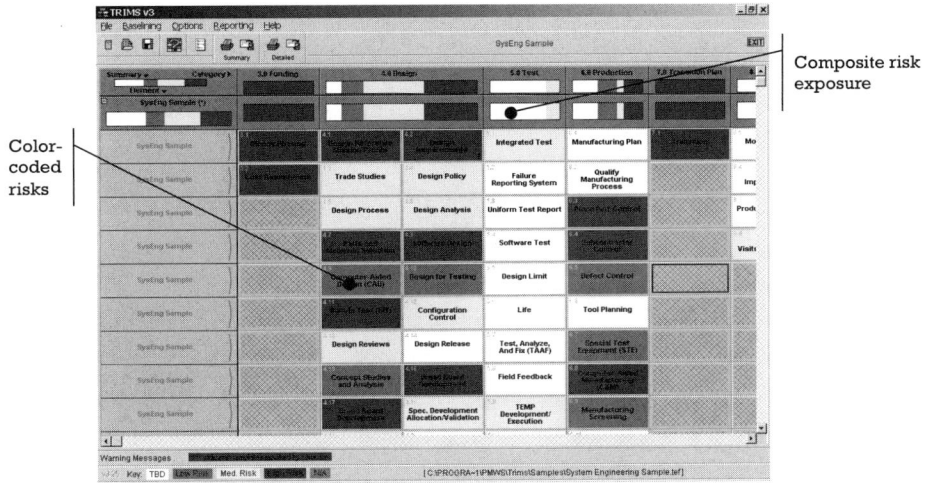

Figure 5.11 Sample risk-management tool: Technical Risk Identification and Mitigation System (TRIMS) risk panel developed by the U.S. Navy for its Best Manufacturing Practices program. TRIMS is based on proven risk models (such as those from the Software Engineering Institute), published practices, and the Navy's best-practices templates and can be applied to all phases of both military and commercial programs.

5.3.8   Action item management

This functionality is used by all members of the PO to raise and track issues that require a formal action or response by other members of the team or the organization. In addition to the normal database functions, this tool must provide automatic notification and escalation capabilities.

5.3.9   Time reporting

Capturing accurate and timely work data is essential in order to establish the true cost and status of a project. Traditionally within the realm of the finance organization and a nuisance to all workers, the time-reporting system plays a critical roll in the entire information system, since work hours are utilized either as raw material or a normalization factor by most of the metrics used in project management.

The time-reporting system should be linked to the project WBS, so that there is a correspondence between the work and the progress reported. Additionally, the system should be able to capture information about—for implementing activity-based costing—the type of activities in which the project staff is involved and—for implementing responsibility accounting—the functional area to which the employee belongs.

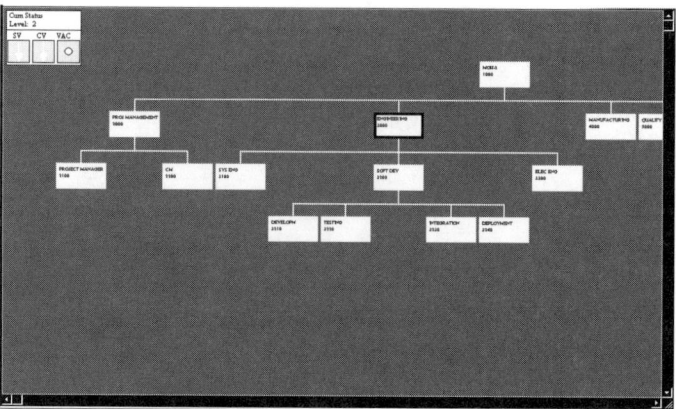

Organizational breakdown structure (OBS)
allows responsibility accounting

(a)

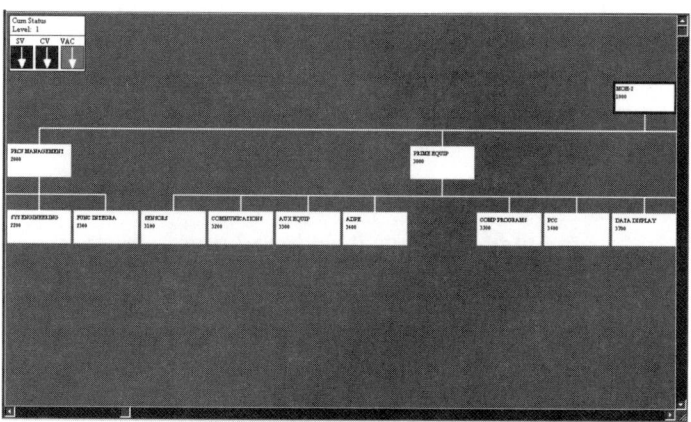

Workbreakdown breakdown structure (WBS)
allows project/product accounting

(b)

Figure 5.12   Organizational and work-breakdown structures in Performance
Analyzer, a tool for the analysis of cost performance reports, cost/schedule status
reports, and contract funds status.

5.4   Commercial tools

In a report in 2000, the Gartner Group [7] segmented the project-portfolio-
management tool market according to the completeness of the vendor
vision and its ability to execute (see Figure 5.14). Since the study was

Cost
accounts
with
variances
and trend
indicators

Cost
Performance
Index (CPI)
Chart

Variance
drivers

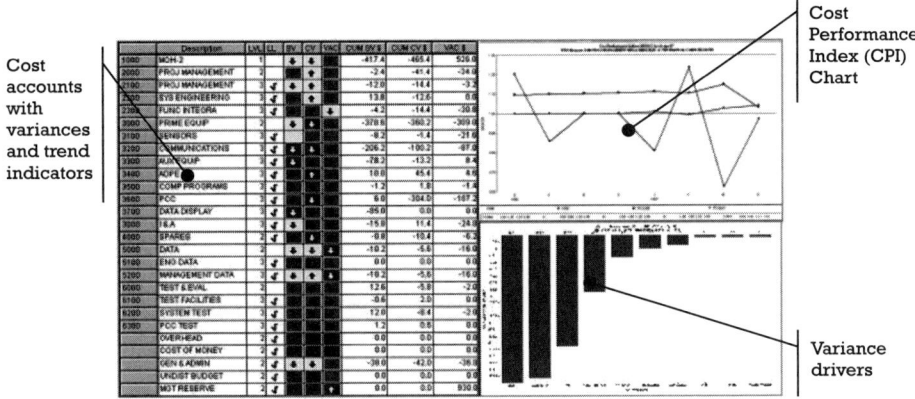

Figure 5.13   Performance Analyzer composite display showing financial performance.

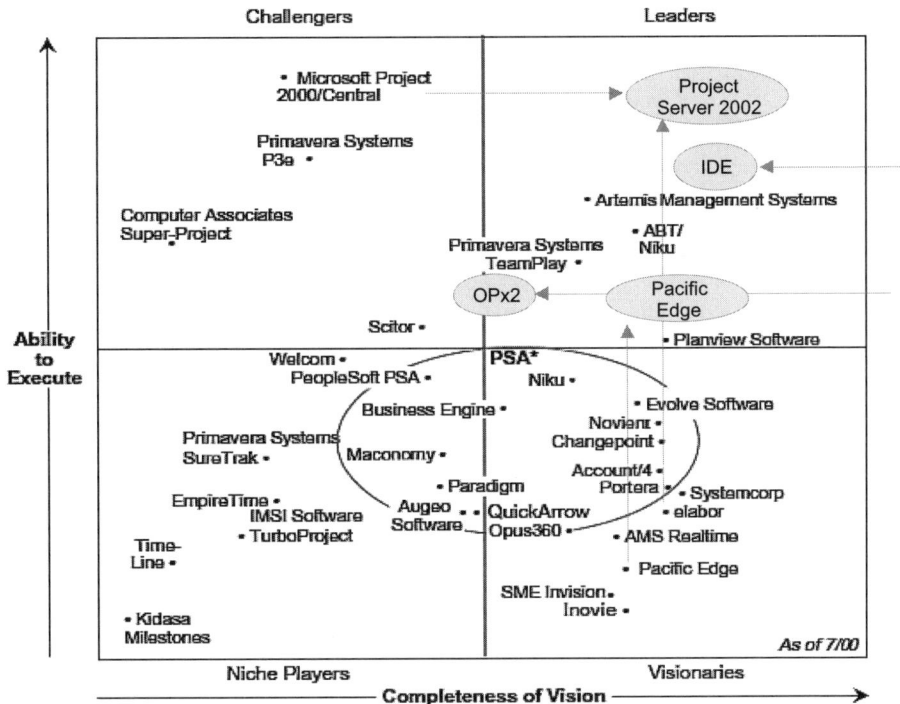

Figure 5.14   Gartner Group's Magic Quadrants. Tools named in the gray ovals were added by the author to reflect product evolution; they do not appear in the original study. (*After:* [7].)

originally published, many changes have occurred: New vendors such as Integrated Development Enterprise (IDE) gained preeminence, others such as PacificEdge demonstrated their ability to deliver, and still others, such as Microsoft and eLabor, joined forces in an effort to dominate the market.

Vendors seem to differentiate their value proposition based on one of two approaches: the platform approach followed by Microsoft Project Server 2002, Primavera P3e, and Planisware OPx2, and the process approach, best represented by IDWeb from IDE, the EDGE for IT from PacificEdge, Plan-View, and Primavera's TeamPlay.

In the platform approach, the vendor provides a powerful application that offers practically limitless possibilities but requires extensive configuration and process work on the part of the customer. In the process approach, the vendor offers an application with less flexibility but one that incorporates the experience of the vendor in a certain area or industry segment, through predefined workflows, project templates, and reports. Table 5.1 provides a necessarily incomplete list of tool vendors.

Table 5.1 Portfolio Management Tool Vendors

Vendor, Product, and Web Site	Description
Microsoft, Project Server 2002, www.microsoft.com	Project Server 2002 is built on the foundation established by eLabor's Enterprise Project. Project Server is a totally new application and has no resemblance to the ill-fated Project Central. The main functionality of Project Server 2002 includes project portfolio planning, competence-based resource management, top-down/roll-up budgeting, and PO and senior management decision support.
	In addition to its extensive native functionality, Project Server 2002 enables the sophisticated user to extend the system flexibility through its Application Programming Interface (API); new custom views can be defined and tailored to the needs of the organization.
PacificEdge, Project Office, www.pacificedge.com	Project Office provides a centralized project repository, resource pool, action item tracking, and document-management capabilities and relies on a seamless integration with Microsoft Project for scheduling. Project Office is organized around four modules: Project Office and Project Office Alerts, through which the user can define business rules and have the tool automatically check them, sending notifications when one or more of these are verified; Project Office Express, which provides time reporting and calendar and group collaboration functions for team members; and the Edge, which provides reporting capabilities at the portfolio level. A fifth module, called the Edge for IT, which despite its name could be used for projects other than IT, provides the knowledge component.
	One remarkable thing about this tool is the simplicity of its interface, which makes it possible for current users of MS Project to be up and running Project Office in almost no time.

Table 5.1   continued

Vendor, Product, and Web Site	Description
IDE, IDWeb, www.ide.com	IDe's founder and chairman, Michael McGrath, is also a founding director of Pittiglio, Rabin, Todd and McGrath (PRTM), a leading management-consulting firm to technology-based companies. He is also a principal contributor to the Product And Cycle-Time Excellence (PACE®) methodology, a well-established stage-gate methodology for new product development, which is manifest in IDWeb philosophy.
	IDWeb consist of a number of modules that together provide a comprehensive set of functionality, including Pipeline Management, Resource Management, Project Planning and Management, Process Management, Financial Management, Partner Management, Time Collection, Partner Management, and Idea Management.
Primavera, P3e, www.primavera.com	Primavera has long been a household name in scheduling and contracting software, especially for the construction industry, and with Primavera Project Planner® for the Enterprise (P3e®), Primavera enters the multiproject planning and control arena.
	A number of products, such as Primavision, Portfolio Analyst, Methodology Manager, Progress Reporter, and the Mobile Manager extend P3e's functionality by providing team member access, time reporting, process management, and portfolio-management capabilities.
Planisware, OPx2, www.planisware.com	Planisware is a French company started in 1996 to commercialize OPx2, a client-server project-management software package, developed from 1991 onwards, with the support of Thomson-CSF, a major aerospace, defense, and electronic conglomerate.
	OPx2 Pro is the central product, providing scheduling, resource management, and cost control functionality. Other modules such as OPx2 TimeCard, OPx2 Server for integration with MS Project, and OPx2 Intranet Server for team member access, complement OPx2 Pro capabilities.
PlanView, PlanView, www.planview.com	PlanView was founded in 1989. Its vision since then has been to create a tool to allow multiple managers to allocate resources from a common pool without overbooking them. In its current version, the tool has exceeded that functionality and now offers portfolio management; service management; resource management; opportunity management; financial management; and collaboration.
Artemis, ViewPoint & Portfolio Director, www.artemisintl.com	The functionality covered by this tool includes planning and scheduling, resource management, timesheet tracking and approval, reminders and notifications, collaboration, and portfolio reporting.
Augeo, Intelligent Planner, www.augeo.com	Augeo is a well-established professional services automation (PSA) provider in the European market. Its value proposition falls in the process arena, with specialized offerings targeting IT departments; R&D departments; pharmaceutical industries; and consultancy groups.
Speed to Market, Concerto, www.speedtomarket.com	Concerto provides multiproject management capabilities based on the critical chain approach to project management. Its functionality includes scheduling, resource management, what-if analysis, and workflow automation.

5.5 Summary

To support an effective PO, any tool or set of tools, whether off-the-shelf or homegrown, must satisfy a few fundamental requirements.

First, tools should adequately reflect the needs of three different types of users: portfolio managers, project managers, and resource managers. Second, they must integrate the different perspectives (finances, resources, and tasks) that make up the portfolio management function, so that when a change in one dimension occurs, it is reflected in the other dimensions as well. Third, they must deliver control information accurately and timely. And fourth, they must have an open architecture that allows the PO to customize them to its own, ever evolving needs.

References

[1]   Goldratt, E., *The Haystack Syndrome: Sifting Information out of the Data Ocean,* Croton-on-Hudson, NY: North River Press, 1991.

[2]   Turner and Speiser, *Program Management and Its Information System Requirements,* 1992.

[3]   Klein, G., *Sources of Power: How People Make Decisions,* Cambridge, MA: MIT Press, 1999.

[4]   Project Control Panel, Software Project Management Network, 1998.

[5]   Miranda, E., "The Use of Reliability Growth Models in Project Management," *9th Int. Symp. on Software Reliability Engineering,* IEEE, Paderborn, Germany, 1998.

[6]   Bleeker, R., "Key Features of Activity-Based Budgeting," *IEEE Engineering Management Review,* Vol. 30, No. 1, First Quarter 2002.

[7]   Light, M., and T. Berg, "The Project Office: Teams, Processes, and Tools," Gartner Group RAS Services, R-11-1530, 2000.

CHAPTER

6

Contents

Balancing the project portfolio

6.1 Introduction

The proper allocation of an organization's finite resources is crucial to its long-term prospect. The most successful organizations are those that have in place a formal project-portfolio-planning process: They allocate staff and budget efficiently, and they quickly terminate projects that do not meet their continuation criteria [1]. A good project portfolio planning process shall be capable of answering the following questions: Of the many projects that the company could pursue, which combination of projects will most closely align with the organization's strategic goals? Which is the best time to execute them? Which will maximize profit? Which will minimize risk? Are there portfolio configurations[1] that perform well on all of these criteria? Are there portfolio configurations that meet all or none of these criteria?

Because of the large number of possible portfolio configurations and the conflicts between the criteria used to select projects, finding the right portfolio configuration is a complex task, which cannot be done using intuition alone; it requires the use of quantitative techniques.

At any given time, an organization has a finite capacity to perform work, and although this capacity could be modified, the process of acquiring or reducing the resources takes time.

---

1. The term *portfolio configuration* is used to denote a unique combination of projects, together with their start and finish dates.

129

Because of this, the organization needs to plan how much work to take in, or if a decision to change the current capacity is made, it must decide when and by how much. Failure to plan leads to paralysis as a result of fire fighting or to inefficiencies in the use of available resources.

The PO facilitates the balancing of the project portfolio by providing senior management, project sponsors, and line managers with the information and tools necessary to make the proper allocation decisions.

Balancing the project portfolio, the most critical part of the planning process, requires the following:

- Calculating the collective requirements of all the projects in the portfolio;
- Calculating the benefits to be derived from the execution of a particular portfolio configuration;
- Identifying resource shortfalls and availability;
- Deciding what to do and when.

Balancing the portfolio is an iterative process (see Figure 6.1) in which the organization selects a certain portfolio configuration, compares the

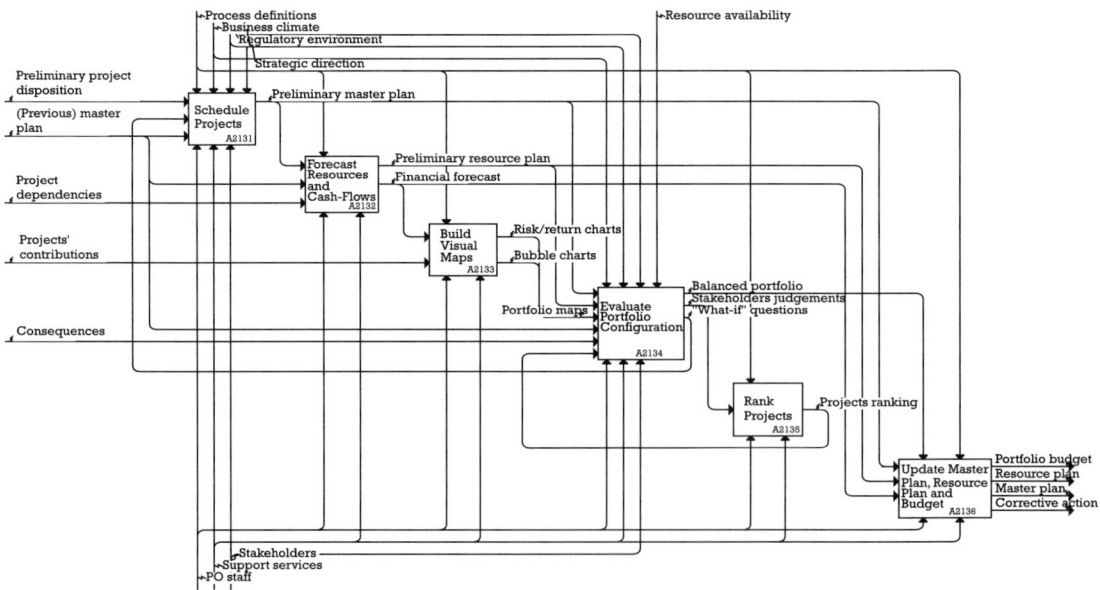

Figure 6.1 Portfolio-balancing process.

workload arising from it to the available capacity, and decides, based on its forecasted contribution, whether to accept the plan as is, to accept it and increase or decrease capacity, or to try a different configuration.

6.2   Project formulation

Although not part of the portfolio-balancing activity, the project formulation process (see Figure 6.2) is the process during which most of the information required for planning and balancing the portfolio is produced. Of particular interest are the following tasks:

- ▸ Producing a rough estimate;
- ▸ Calculating contingency allowances (risk and opportunity management);
- ▸ Evaluating individual projects' contribution to benefits (project concept definition);
- ▸ Establishing project dependencies (updating dependency matrix).

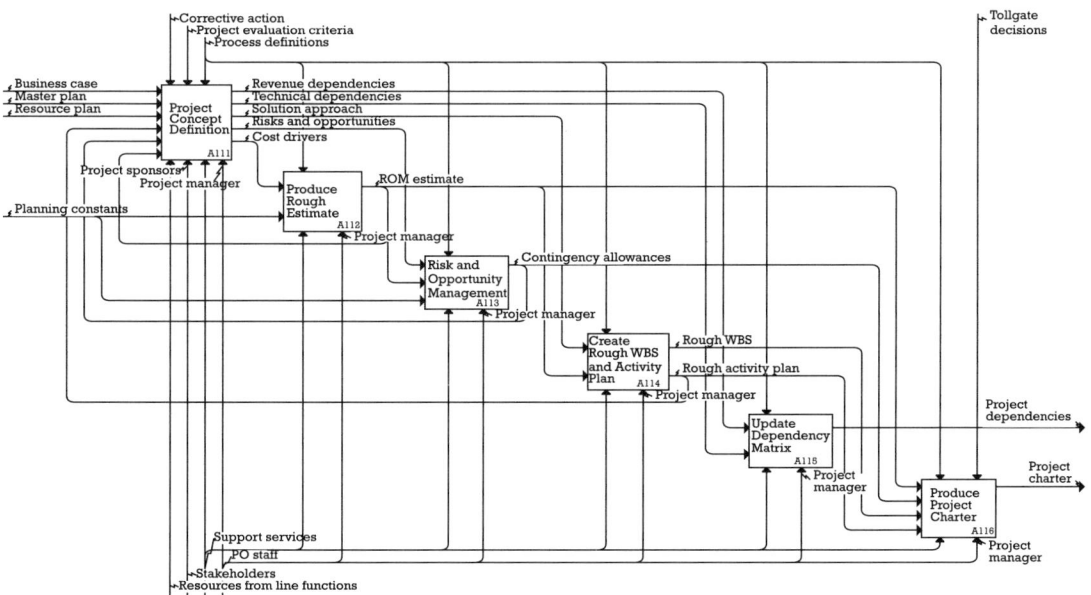

Figure 6.2   Project formulation process.

6.2.1   Producing a rough estimate

Cost, effort, and schedule estimates should be developed for the proposed project scope from relevant cost drivers. For example, functional size and complexity would be considered in the case of software; estimated number of devices would be taken into account in hardware projects; and therapeutic category and the number of chemical reactions necessary to synthesize a molecule [2] would be required for the pharmaceutical industry.

Once the initial cost, effort, and schedule estimates have been developed, before committing to the estimate, the following questions should be answered:

- Does the estimate make sense?
- Are estimated schedules, costs, and effort consistent with prior experience?
- Do the estimated effort, cost, and schedule meet programmatic requirements?
- Are required productivity levels reasonable?
- Have all relevant costs drivers been included?

The analysis of the estimate serves four purposes: to ensure that the estimate is thoroughly understood, to ensure that the estimate is as accurate as possible; to provide a baseline upon which to evaluate the project benefits, and to conduct risk analyses. Typical methods used in the development of rough estimates, as described in Table 6.1, are expert judgment, analogy, and parametric models.

Table 6.1   Estimation Methods

Estimation Approach	Description	Advantages	Limitations
Analogy	Compare project with past similar projects	Estimates are based on actual experience	Truly similar projects must exist
Expert judgment	Consult with one or more experts	Little or no historical data is needed; good for new or unique projects	Experts tend to be biased; knowledge level is sometimes questionable
Parametric models	Perform overall estimate using design parameters and mathematical algorithms	Models are usually fast and easy to use, and useful early in a program; they are also objective and repeatable	Models can be inaccurate if not properly calibrated and validated; historical data used for calibration may not be relevant to new programs

Whatever the technique used in developing them, estimates are contingent on the many assumptions upon which they rely and over which the project manager has little or no control. Moreover, estimates could reflect bias on the part of those producing them. Because of uncertainty surrounding the estimated cost, effort, and schedule, three-point estimates (best case, worst case, and most-likely scenario) from which a probability distribution could be derived (see Figures 6.3 and 6.4) are a better alternative to single-number estimates.

The schedule risk of any project, as illustrated in Figure 6.5, manifests itself as a project with the same "most likely" completion date, but farther to the right, a worst-case completion date. In statistical terms, this variation in the spread of completion dates is captured by the standard deviation of the distribution. Similar considerations could be made with respect to the effort and cost risks of a project.

Since a project is a unique happening, it is impossible to know its real effort and duration probability distributions, and in consequence, any distribution chosen would only approximate the true, but unknown,

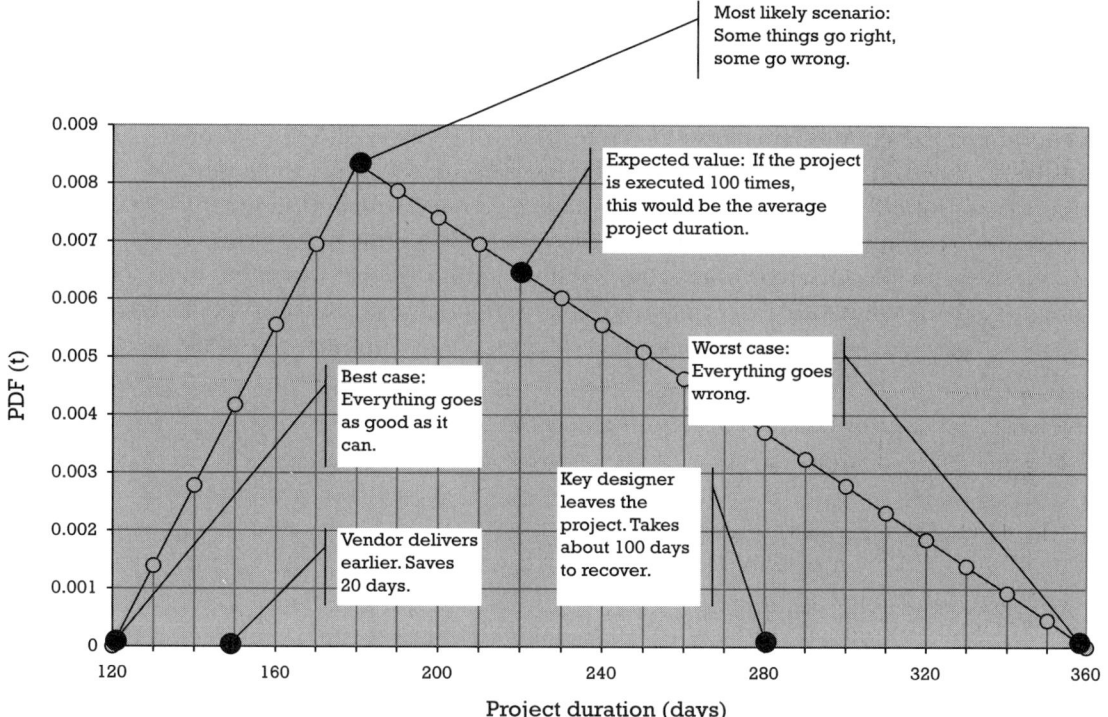

Figure 6.3   Three-point estimate.

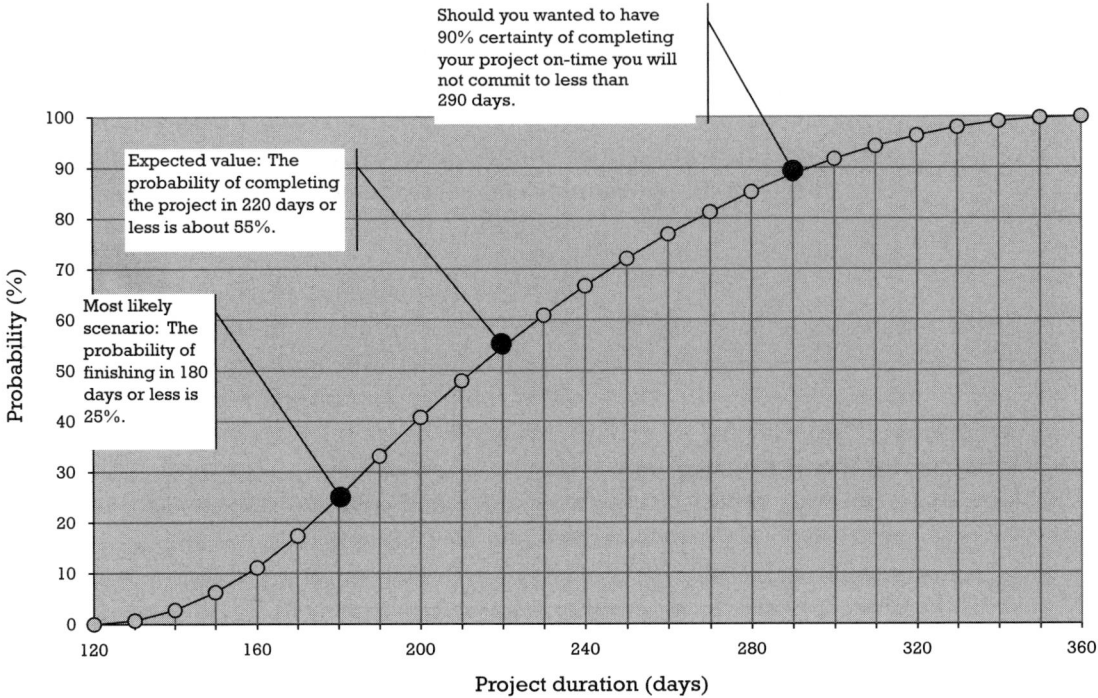

Figure 6.4 Cumulative probability distribution.

distribution. Given this fact, the use of a triangular distribution to model project risks is a sensible choice [3] because its parameters are easy to understand, its shape is consistent with project managers' and sponsors' expectations, and its probabilities are easy to compute, even without a pocket calculator. Assuming a triangular distribution, the expected effort and project duration and its standard deviations are calculated using the following expressions:

$$ExpectedProjectEffort = \frac{a + b + c}{3}$$

$$\sigma(ProjectEffort) = \sqrt{\frac{a^2 + b^2 + c^2 - ab - ac - bc}{18}}$$

where a is the least effort at which the project can be completed, b the most likely, and c the worst case. $\sigma(ProjectEffort)$ denotes the standard deviation of the required effort. Similarly, the expected project duration can be calculated by the expression

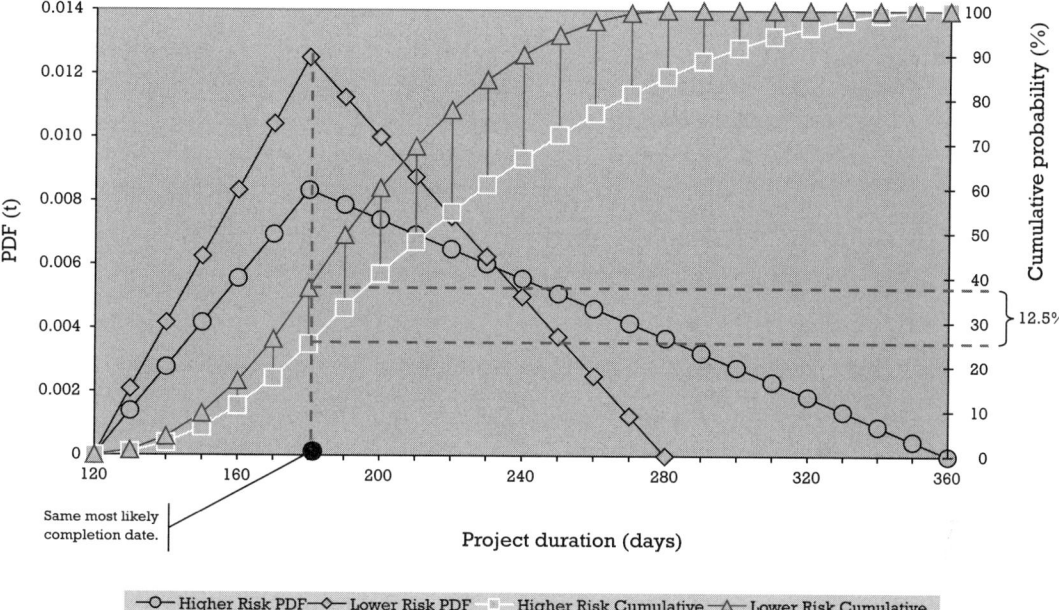

Figure 6.5 Project schedule risk.

$$ExpectedProjectDuration = \frac{a + b + c}{3}$$

$$\sigma(ProjectDuration) = \sqrt{\frac{a^2 + b^2 + c^2 - ab - ac - bc}{18}}$$

where a is the shortest time in which the project can be completed, b the most likely, and c the worst case. $\sigma(ProjectDuration)$ denotes the standard deviation of the required effort.

The probability of a project finishing on or before a certain time t, or requiring less than a certain effort, $F(t)$, could be calculated as follows:

If $t < a$ then
$$F(t) = 0$$
if $a \leq t < b$ then
$$F(t) = 1 - \frac{(t - a)^2}{(b - a)(c - a)}$$
if $b \leq t < c$ then

$$F(t) = 1 - \frac{(c-t)^2}{(c-a)(c-b)}$$

if $c \leq t$ then
$$F(t) = 1$$

Choosing a 50% certainty in the completion date of the projects means that in the long run, one of every two projects will finish late. Choosing a 75% probability will reduce this number to one in four.

6.2.2 Calculating contingency allowances

Most street-smart project managers will hide funding in their budgets for unknown contingencies, and most savvy managers and sponsors, being aware of this, will cut the budgets back. The end result of this silly game is that there is no or little visibility with respect to these funds, how much has been allocated, who controls them, and when and on what they should be spent.

Clearly spelling out project risks and using an insurance-like mechanism for the management of contingency funds will result in the following benefits to the organization:

▸ Better profitability decisions, as the risks posed by a project are taken into consideration in the business decision;

▸ Reduced cost of capital, as the organization's budgets are based on a risk-spreading policy rather than on worst-case scenarios;

▸ Reduced capital needs, as the "budget allocated is budget spent" syndrome is avoided;

▸ Better anticipation of business needs on the part of those with profit-and-loss responsibilities, who will attempt to avoid being hit with an insurance premium that will go directly against their bottom line.

According to the *Encyclopaedia Britannica*, insurance is "a contract for reducing losses from accidents incurred by an individual party through a distribution of the risk of such losses among a number of parties." In other words, while the destruction of an automobile in a traffic accident imposes a heavy financial loss on an individual, one such loss is of relatively small consequence to an insurer who is collecting sufficient premiums on a large number of automobiles.

While the cost of ensuring a single project against all possible risks and uncertainties would be prohibitive, major risk categories, such as schedule, technology, and organizational maturity, could be singled out, and the project required to include in its budget an insurance premium as part of the cost of doing business. As an example, a project with a highly compressed schedule—a high risk factor—will have a higher insurance premium than a project with a more relaxed one. Taking this additional cost into account could lead to a totally different decision from that made without factoring such risk into the business case.

As an example (see Figure 6.6) let us look at a project with a very aggressive schedule requested by marketing to fit the window of opportunity. The targeted completion date for the project is 150 days, and this date has a probability of being met of around 6.2%. The probability of being late is therefore over 93%. Assuming that the policy of the organization is to have its projects scheduled so that its probability of being late is at most 25%, this would leave the project with a 68.75% (75% – 6.25%) probability of being late above the accepted risk. The question then becomes, should the project need to expend extra money to keep to the promised schedule, who is responsible for the additional funding?

If the organization charges all of the risk to the project, it might price the product out of competition. If it decides not to charge anything, it is likely that the organization will have to complete the project at a loss. A fair

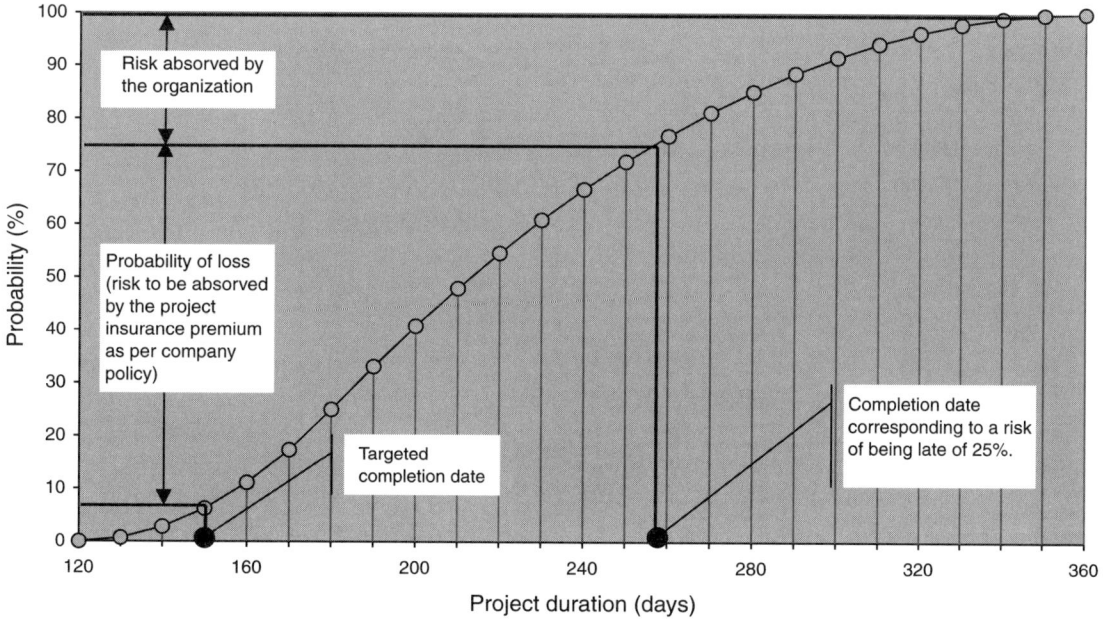

Figure 6.6   Schedule-risk calculations.

valuation of this risk could be established using the insurance underwriter's most basic equation:

$$InsurancePremium = \frac{ProbabilityOfLoss \times MagnitudeOfLoss}{[1 - (ExpenseRatio + ProfitRatio)]}$$

The probability of loss in this equation is the probability of not being on time as calculated above. The expense and profit ratio could be replaced by the organization's cost of borrowing. The magnitude of the loss is the extra amount of money necessary to keep to the schedule when this has been underestimated. The diagram in Figure 6.7 shows that in order to recover from an underestimation, it is necessary to use more resources than would be required if the amount underestimated would have been included in the original plan. The extra resources are needed to compensate for the following:

• The time it will take to find them;
• The time it will take to bring them up to speed;
• The time taken away from other staff in order to bring the new resources up to speed.

The equations below, derived from the geometry of Figure 6.7, are used to calculate the recovery cost as follows:

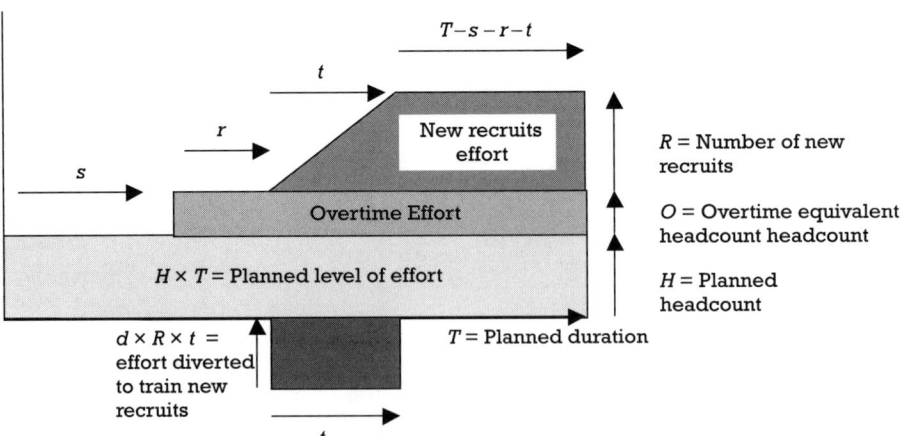

Figure 6.7 Cost of recovery. (*Source:* [3].)

$$CostOfRecovery = R(T - r - s) \times m + O(T - s) \times n$$

$$R = \frac{(A - HT - O(T - s))}{(T - s - r - 0.5t - dt)}$$

where

$A =$ effort required to complete the project at the risk level decided by the organization

$H =$ head count allocated as per original plan

$T =$ planned duration

$O =$ equivalent head count provided by overtime

$R =$ number of people to be added to the project to attempt a recovery

$s =$ time it takes to come to the conclusion that the project is going to be late

$r =$ time it takes to incorporate new people into the project

$t =$ time it takes for a new person to become fully productive

$d =$ proportion of effort that an original member of the team devotes to each newcomer

$m =$ regular cost per resource per unit of time

$n =$ overtime cost per resource per unit of time

Table 6.2 develops a numeric example of the calculation process and Figures 6.8 and 6.9 plot the economics of project insurance.

The insurance premium is "payable" to the PO, which is responsible for the administration of funds. The PO is then able to isolate the organization from fluctuations in individual projects' budgets, which will therefore start to see its projects come in on budget.

Ideas similar to these can be found in the risk analysis and cost management model developed at Lockheed Martin Missiles and Space Company, Sunnyvale, California [4], the probabilistic analyses of technical, schedule and cost risks performed for major projects at the Los Alamos National Laboratory (LANL) [5], and the planning of contingency funds at Compaq [6].

6.2.3   Evaluating individual projects' contribution to benefits (project concept definition)

Evaluating the contribution of individual projects is an intrinsic part of the project-concept-definition activity. Individual projects might contribute to the organization's goals along a number of dimensions. For for-profit

Table 6.2   Project Insurance Calculation

Step	Input and Process	Result
1	Assume the following conditions: Organization policy requires that all projects be planned with a 75% level of certainty Targeted project duration (T) = 150 days Planned head count (H) = 10 people Equivalent overtime head count (O) = 2 people Average time to detect a slowdown (s) = T/2 = 75 days Lead time to get new people transferred into the project (r) = 20 days Ramp-up time (t) = 20 days Percent of effort diverted to help newcomers (d) = 0.2 m = \$800/person day n = \$1200/person day Cost of borrowing = 10%	
2	Calculate the required effort (A): From Figure 6.7, the duration of a project with this level of risk, according to company policy, must be 258 days in order to have a 75% chance of not exceeding the project budget. $A = PolicyRequestedDuration \times H$	$A = 258\,days \times 10\,persons = 2{,}580\,person-day$
3	Calculate the cost of recovery: $R = \dfrac{(A - HT - O(T-s))}{(T - s - r - 0.5t - dt)}$ $CostOfRecovery = R(T - r - s) \times m + O(T-s) \times n$	$R = \dfrac{(2{,}580 - 1{,}500 - 2 \times (150-75))\,person-days}{(150 - 75 - 20 - 0.5 \times 20 - 0.2 \times 20)} = 22.68\,persons$ $CostOfRecovery = 22.68 \times (150 - 20 - 75) \times 800 + 2x(150-75) \times 1{,}200$ $= \$1{,}178{,}048$
4	Calculate the insurance premium for the project: $InsurancePremium = \dfrac{ProbablilityOfLoss \times RecoveryCost}{[1 - CostOfBorrowing]}$	$ProbablilityOfLoss = 0.75 - 0.0625 = 0.6875$ $InsurancePremium = \dfrac{0.6875 \times 1{,}178{,}048}{[1 - 0.10]} = \$899{,}897$

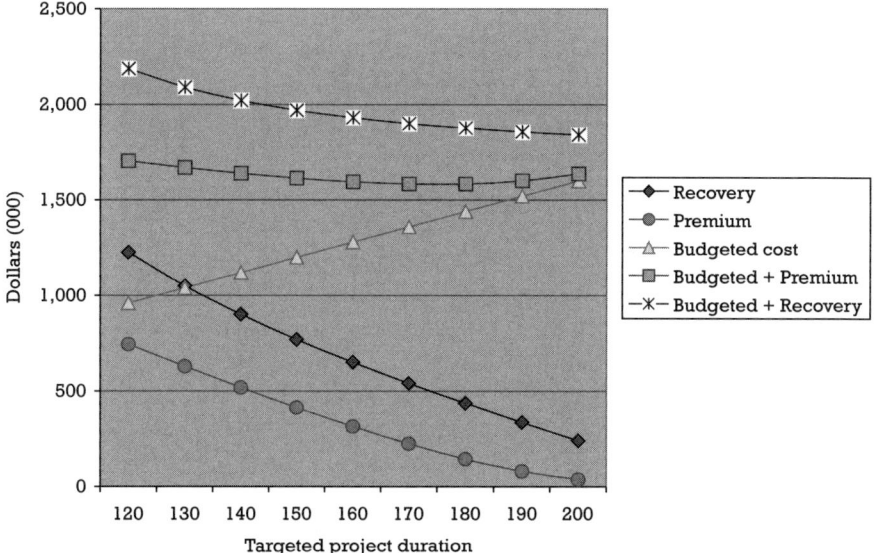

Figure 6.8   Economics of project insurance—acceptable risk level set at the expected project completion date.

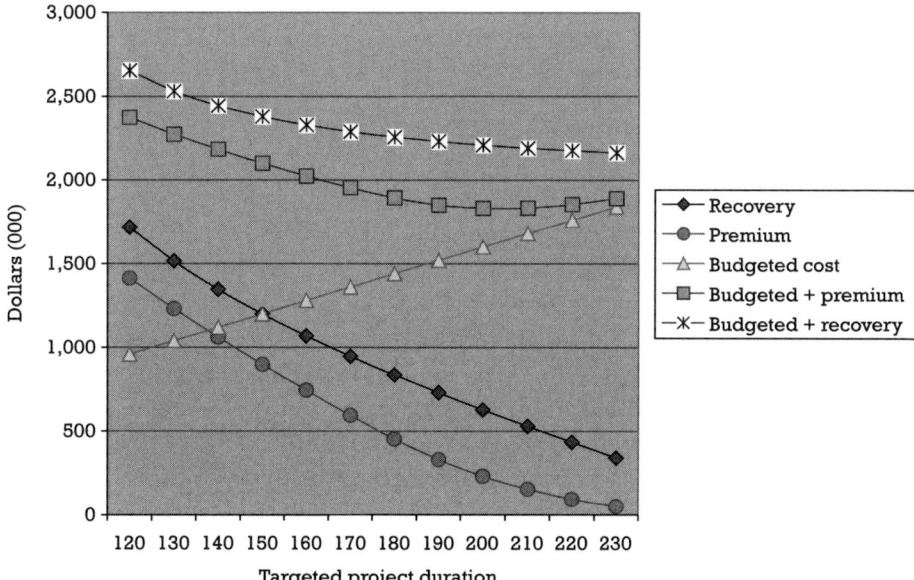

Figure 6.9   Economics of project insurance—acceptable risk level set at the project completion date with a 75% probability of being met.

organizations, financial success is obviously a very important dimension, but by no means the only one. Common dimensions to be evaluated when judging the merits of a project include strategic position, probability of technical success, probability of commercial success, sociopolitical and regulatory consequences, costs, rewards, and nature of work.

The fact that a project scores low with respect to one dimension does not mean that the project should be automatically disqualified. A project could bring benefits to the organization by allowing it to develop a new competency or by providing entry into a new business segment, and in such cases, projects might be undertaken at a loss with the hope of harvesting larger benefits later.

The importance of the evaluation process resides as much on the final numeric results it produces as on the review process it forces upon the decision makers. The consistent application of the evaluation criteria is necessary to arrive at a thorough and fair comparison among the projects, and that is why it is necessary to define what those criteria are.

The strategic positioning dimension (see Table 6.3) measures where the project fits within the overall organizational strategy. The execution of the

Table 6.3   Strategic Position

Evaluation Criteria	Ranking			
	Unfavorable			Favorable
Congruence	No fit with defined strategy; success will not bring about a new strategy	Marginal contribution to defined strategy	Direct fit and good contribution toward one element of the strategy	Strong fit toward several elements of the strategy; success could open up new opportunities
Impact (if project dropped or unsuccessful)	Moderate competitive and financial losses	Significant competitive and financial losses	Would lose current position; could take years to recover	Future of the business depends on this project
Synergy with other projects	One-of-a-kind, dead-end work	Slight change that will lead to repeated business	Continuation work has already been announced	Required as part of a multiyear program
Learning opportunities	No skills enhancement; have done it before	Adds new capabilities within current business		Develops new capabilities in different business area

project could strengthen the competitive position of the organization along the lines of a defined strategy, or it could be a totally disruptive initiative whose success will allow the organization to enter into a new, previously un-thought-of business segment by developing a new competence.

The technical success probability measures the projects with respect to their likelihood of delivering what they are supposed to deliver. Obviously, other things being equal, the higher the probability of success of a project the better. This does not mean, however, that projects with a low probability of success should not be undertaken; it does mean, though, that for an organization to embark on such projects, the risks need to be understood and the rewards need to be consistent with the exposure incurred.

Although technical risk taxonomies are industry specific, most project failures could be traced to one or more of the factors depicted in Table 6.4.

A technically successful project might still be a commercial failure should the organization fail to anticipate the size of the market, the availability of suppliers, or the strength of the competition. The probability of commercial success (see Table 6.5) measures the likelihood of the project results to be successful in the marketplace.

The projects an organization undertakes can have an impact on the community at large, or they might be imposed upon the organization by

Table 6.4 Probability of Technical Success

Evaluation Criteria	Ranking			
	Unfavorable			**Favorable**
Technology readiness level	Large gap between practice and target; must invent.	Technology demonstrated in the lab, but not in actual application.	Technology validated in a relevant environment.	Incremental improvement of existing technology.
Complexity	The final product involves over 5,000 modules, components, or assembly steps.	The final product involves less than 5,000 modules, components, or assembly steps.	The final product involves less than 200 modules, components, or assembly steps.	The final product involves less than 50 modules, components, or assembly steps.
	Quality, price, and performance targets could be achieved only by considerable optimization work.	Quality, price, and performance targets could be achieved only by considerable optimization work.	Quality, price, and performance targets could be achieved by paying attention to details.	Easy to meet quality, price, and performance targets.
Availability of people, facilities, and time	No appropriate people/facilities, must hire/build.	Acknowledged shortage in key areas.	Resources available but in demand.	People/facilities available immediately; agreed schedule.
	Impossible schedule.	Challenging schedule.		

Table 6.5   Probability of Commercial Success

Evaluation Criteria	Ranking			
	Unfavorable			**Favorable**
Competition	Many similar offerings; competing in price	Many similar offerings; competing in quality	Similar offers, distinctive value proposition	Unique solution
Market size	Single client	Multiple clients		Market
Product life-cycle stage	Declining	Mature	Growth	Embryonic

society through regulations and norms with which the organization, or the products it produces, must comply. The sociopolitical dimension (see Table 6.6) measures the projects with respect to the regulatory situation and with respect to the personal and environmental consequences that might arise from the execution of the project or the use of its results.

The rewards dimension measures the payoff to be derived directly from the execution of the project. In this dimension (see Table 6.7), we include the techniques for the economic analysis of the project, such as net present value (NPV), internal rate of return (IRR), return on investment (ROI), and real options valuation (ROV).

In addition to the cost and schedule estimates required for the economic evaluation of the project, it is necessary to evaluate the magnitude of the effort required in relation to the resources of the organization. A million-dollar project could be a very small or very large project, depending on the size of the organization. The cost dimension of a project (see Table 6.8) measures this relationship.

The nature of the work is an important but often neglected dimension, especially among successful organizations. The nature-of-work measurement classifies projects according to whether they support current initiatives

Table 6.6   Sociopolitical and Regulatory Consequences

Evaluation Criteria	Ranking			
	Unfavorable			**Favorable**
Regulatory environment	Does not meet current regulations	Extensive qualification needed to satisfy current regulations	Satisfies current regulations	Sets the standard Exceeds current regulations
Environmental hazards	Increases emissions subject to regulation	Reduces emissions subject to regulation	Replaces current emissions requirement with one less hazardous	Eliminates regulated emissions

Table 6.7   Rewards

Evaluation Criteria	Ranking			
	Unfavorable			**Favorable**
Revenue	The project is done at a loss	Project breaks even	Attractive revenues	Very attractive revenues
NPV/IRR/ROI/ROV	Negative	Does not meet the organization's hurdle rate	Meets the organization's hurdle rate	Exceeds the organization's hurdle rate
Time to break even	More than five times the project duration		Double the project duration	Project duration

Table 6.8   Cost

Evaluation Criteria	Ranking			
	Unfavorable			**Favorable**
Resources	Massively expensive Prevents the organization from devoting resources to any other project	Very expensive Requires major sacrifices; the organization is forced to cancel or opt out of other interesting endeavors	Moderately expensive The organization is able to fund other initiatives	Affordable project
Time	5 years or more		2 years or less	1 year or less

or products or formulate the basis upon which the next generation of products will be built. The purpose of doing this is to align project work with product strategy: Are we taking care of the cash cows? Do we have enough stars? Are we spending money on any dogs? See Table 6.9.

6.2.4   Establishing project dependencies (updating dependency matrix)

Projects would typically build on top of the knowledge, features, and market share developed by preceding projects. These dependencies manifest themselves in the assumptions that justify plans and business cases. Obviously, in the event of the cancellation, delay, or downsizing of any of these projects, those plans and business cases would have to reviewed; even better, before deciding on a cancellation or downsizing, we should take a look at the consequences of such a decision across the entire portfolio. Failure to consider this ripple effect could easily lead to a situation worse than the one we were trying to solve (see Figure 6.10).

Table 6.9 Nature of Work

Nature of work	In support of: Product Line 1	Product Line 2	Product Line 3	...	Product Line n
Technology development					
New platform					
Platform extension					
New derivative product					
Product extension					
Branding					
Fix					

By making the dependencies explicit and recording them on the dependency matrix introduced in Chapter 3 or in a special matrix called the propagation matrix, it would be possible to calculate the effects of any delay or cancellation decision across the project portfolio [7].

The concept of the propagation matrix and its use to spread the effect of a decision is explained in Section 6.3.5.

6.3   Portfolio balancing

The level of utilization an organization can sustain is determined not only by the amount of resources it possesses, but also by the nature of its work. The more unpredictability in the workload and the longer the duration of the projects it undertakes, the more white space or slack the organization will need to consistently deliver on its commitments.

6.3.1   Forecasting resource needs

Once the project's expected effort has been calculated, it is necessary to break it down according to competence types and spread it over the expected project duration. Since at this stage detailed plans for the project might not yet exist, the breakdown and the spreading of the effort would have to be based on project profiles, rather than on actual allocations.

Each project profile would specify the following:

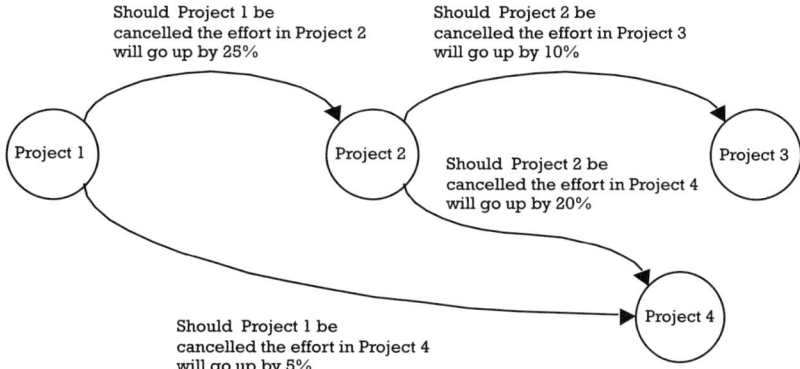

Assuming efforts of 10, 20, 30 and 40 thousand hours for Projects 1, 2, 3 and 4. Should Project 1 be cancelled, the distribution of effort would be as follow:

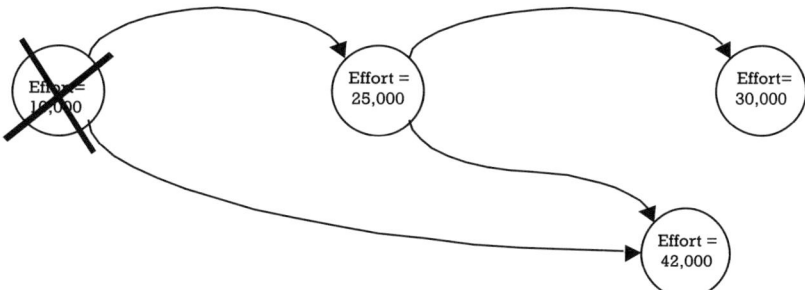

Should Project 2 be cancelled, what would the distribution of effort be?

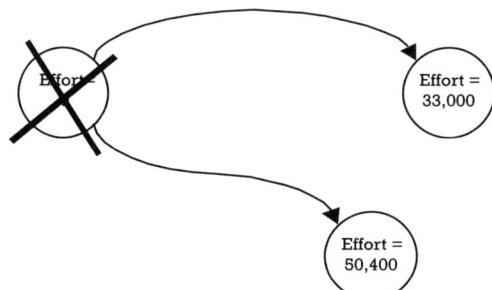

So the total savings after canceling Projects 1 and 2 equals 16,600, not 30,000 man-hours

Figure 6.10 Project cancellation and the propagation of consequences.

> • The *EffortBreakdownRatio$_{pc}$* for competence area c in projects of type p. With $\sum\limits_{c=1}^{N} EffortBreakdownRatio_{pc} = 1$.

> • The *SpreadRatio$_{pct}$* for projects of type p of the apportioned efforts to competence c at relative time t. Again $\sum\limits_{t=1}^{M} SpreadRatio_{pct} = 1$.

An organization will typically have several project profiles, one for each class of projects, such as new platform development, product extension, research, and maintenance. The monthly or quarterly demand for each competence for a given project is calculated using the following expression:

$$ProjectDemand_{ict} = ExpectedProjectEffort_i \times EffortBreakdownRatio_{pc} \times SpreadRatio_{pct}$$

See Table 6.10 for a numerical example of the utilization of techniques used to calculate the resource needs for a typical software development project.

Once the time-phased needs of the projects have been calculated, it is time to aggregate the individual requirements into the master plan. The aggregated demand would be

$$PortfolioDemand_{ct} = \sum_{i \in Portfolio} ProjectDemand_{ict}$$

Since the projects in the portfolio have different objectives, teams, and project managers, it can be assumed that the actual time and effort required by each is largely independent of the others; in consequence, the standard deviation of the portfolio workload could be calculated as the square root of the sum of the squares of the standard deviations of each individual project. See below.

$$\sigma(PortfolioDemand_{ct}) = \sqrt{\sum_{i \in Portfolio} \sigma(ProjectDemand_{ict})^2}$$

The contingency to be added to *PortfolioDemand$_{ct}$* to avoid the risk of exceeding the expected workload with a predetermined probability p can be calculated using the expression below[2]:

---

2. The formula for k is derived from a one-tailed version of Chebyshev's inequality, which states that

$$P(x - E(x) \geq k\sigma_x) \leq \frac{1}{(1+k^2)} \ [8].$$

Table 6.10 Forecasting Resource Needs

Step	Inputs and Process	Result

1	The expected duration and its standard deviation are calculated using a three point estimate. Assume that:
	  Minimum project duration is 10 months
	  Most likely project duration is 12 months
	  Maximum project duration is 18 months
		$ExpectedProjectDuration = \dfrac{10+12+18}{3} = 13.3$
		$\sigma(ProjectDuration) =$
		$\sqrt{\dfrac{100+144+324-120-180-216}{18}} = 1.7$

2	The expected effort and its standard deviation are calculated using a three point estimate. Assume that:
	  Minimum project effort is 15,000 man hours
	  Most likely project effort is 20,000 man hours
	  Maximum project effort is 25,000 man hours
		$ExpectedProjectEffort = \dfrac{15+20+25}{3} = 20$
		$\sigma(ProjectEffort) =$
		$\sqrt{\dfrac{225+400+625-300-375-500}{18}} = 2.04$

3	The expected effort calculated in Step 2 is allocated to the different disciplines, competences or resource types taking part in the execution of the project. The project profile for this type of project specifies the planning constants to be used. | Competence	Relative Allocation
---	---
Project Management	10%
Design	20%
Coding	30%
Integration & Testing	40%
	100%			Competence	Effort in '000 of man-hours
---	---
Project Management	2
Design	4
Coding	6
Integration & Testing	8
	20	

$$PortfolioContingency_{ct} = k \times \sigma(PortfolioDemand_{ct})$$

$$k = \sqrt{\frac{1}{p} - 1}$$

Or if the number of concurrent projects in the portfolio is greater than fifteen,[3] this can be done by looking at a table of normal probabilities.

---

3. We assume that the distribution of the sum of the projects' demands tend to a normal distribution via the central limit theorem. More stringent conditions could be imposed or other distributions deemed more appropriate under special circumstances, but this will come at the expense of formulaitons that are more complicated. Therefore, a balance must be struck between accuracy and practicality.

Table 6.10 (continued)

Step	Inputs and Process	Result
4	The effort for each discipline is spread over the duration of the project by multiplying the allocated man hours, see Step 4, by the constants below. Different types of projects are likely to have different constants. The constants form part of the project profile for that type of project. The normalized time is converted to project time by dividing the expected time calculated in Step 1 by the number of intervals, in this case 12, and multiplying it by the interval number.	(tables and chart below)

Project Time

Competence	1.10	2.20	3.30	4.40	5.50	6.60	7.70	8.80	9.90	11.0	12.1	13.2	Sum
PM	0.17	0.17	0.17	0.17	0.17	0.17	0.17	0.17	0.17	0.17	0.17	0.17	2.00
Design	0.20	0.40	0.60	0.60	0.60	0.40	0.20	0.20	0.20	0.20	0.20	0.20	4.00
Coding	0.27	0.27	0.27	0.55	0.82	0.82	0.82	0.82	0.55	0.27	0.27	0.27	6.00
I&T	0.36	0.36	0.36	0.36	0.36	0.36	0.73	1.09	1.09	1.09	1.09	0.73	8.00

Normalized Time

Competence	1	2	3	4	5	6	7	8	9	10	11	12	Sum
PM	0.08	0.08	0.08	0.08	0.08	0.08	0.08	0.08	0.08	0.08	0.08	0.08	1.00
Design	0.05	0.10	0.15	0.15	0.15	0.10	0.05	0.05	0.05	0.05	0.05	0.05	1.00
Coding	0.05	0.05	0.05	0.09	0.14	0.14	0.14	0.14	0.09	0.05	0.05	0.05	1.00
I&T	0.05	0.05	0.05	0.05	0.05	0.05	0.09	0.14	0.14	0.14	0.14	0.09	1.00

Project workload

Step	Inputs and Process	Result
5	The result is ready to be aggregated with other projects in the portfolio.	

$$PortfolioContingency_{ct} = k \times \sigma(PortfolioDemand_{ct})$$
$$z = G^{-1}_{N0,1}(ProbabilityOfFinishingLate)$$

Table 6.11 shows the amount of contingency to be added to the expected portfolio demand so that the probability of exceeding that amount will be lower than a prescribed value.

Once the workload and the contingency are calculated, these numbers must be transformed into head count by dividing the aggregated number of hours required by the effective number of work hours per month. The conversion factor from man-hours to head count can be easily calculated by subtracting from the nominal number of work hours, the average number of training hours, vacation, and sick leave. More mature organizations can include other factors, such as recruiting lead time, learning curves, and

Table 6.11   Contingency Table

Probability of Exceeding Planned Resources		Contingency To Be Added for Each Resource Type	
		Fewer than 15 concurrent projects	15 or more concurrent projects
Higher risk tolerance	25%	$1.73 \times \sigma(PortfolioDemand_{ct})$	$0.68 \times \sigma(PortfolioDemand_{ct})$
	20%	$2.00 \times \sigma(PortfolioDemand_{ct})$	$0.84 \times \sigma(PortfolioDemand_{ct})$
	15%	$2.38 \times \sigma(PortfolioDemand_{ct})$	$1.03 \times \sigma(PortfolioDemand_{ct})$
Lower risk tolerance	10%	$3.00 \times \sigma(PortfolioDemand_{ct})$	$1.28 \times \sigma(PortfolioDemand_{ct})$

turnover rates; however, it is important to remember that for resource planning purposes, the consistent use of the numbers is as important, if not more important, as their accuracy. This is because if different departments use different definitions to plan and account for time, the results will be irremediably meaningless.

The contingency or slack added to the plans is not idle time. It is the time that the organization would use to do its training, its process improvement, and other like activities that could be postponed without too much harm should the need arise. The use of contingency time, however, must be closely scrutinized so that it is not used to hide or compensate for deficiencies in the planning and execution of the projects at the expense of these nonurgent, but nonetheless critical, activities for the long-term survival of the organization.

6.3.2   Forecasting revenues

Once the product or business managers have estimated the revenues to be generated by a given project, the process of spreading them over the life span of the product and aggregating across the portfolio is, with two exceptions, identical to the one used for spreading the resource demands of a project.

The first difference between the resource demands and the revenues stems from the time value of money. One hour of work today and one hour of work tomorrow represent exactly the same value. This is not the case with money, which is affected by inflation and by the cost of capital. In practical terms, this means that any amounts spread over time must be adjusted by inflation and by the cost of capital for any comparison to be meaningful. The formula to convert any future amounts to today's value is

$$PresentValue = \frac{Amount}{(1+i)^t}$$

where

i = discount rate

t = period in which the *Amount* takes place

The second difference is that the assumption of independence used in the calculation of the standard deviation of the portfolio demand might or might not hold for the projects' revenues. As an example in which the revenues of two projects are not independent, take the sales generated by a project that delivers an update to a product developed by a previous project. Obviously, the revenues to be generated by the update will depend largely on the sales of the original product. If the original product was sold in large quantities, the update will likely sell in large quantities as well. Likewise, if the original product did not sell well, the update is not likely to sell well either.

In the example above, the revenues to be generated by both projects are correlated. Although the aggregated value of the revenues is calculated as before, the standard deviation is calculated using the formula below, or by means of Monte Carlo simulations.

$$\sigma(PortfolioRevenues_t) =$$

$$\sqrt{\sum_{i=1}^{n} \sigma(ProjectRevenue_{it})^2 + 2\sum_{i=1}^{n-1}\sum_{j=i+1}^{n} \rho_{ij}\sigma(ProjectRevenue_{it})\sigma(ProjectRevenue_{jt})}$$

where $\rho_{ij} \in [-1,1]$ = the extent of the relationship between projects i and j; -1 means a perfectly inverse relationship; 1 a perfectly direct relationship; 0 no relationship at all; and the rest of the numbers something in between. The value of ρ_{ij} must be "guessed" by the product manager.

6.3.3 Eliminating less valuable alternatives

Whenever a decision to scale back, terminate, or not pursue a particular project is about to be made, in addition to any direct losses or gains arising from it, it is necessary to consider the consequences of the decision across the entire portfolio.

A propagation matrix[4] (see Table 6.12) is a square matrix that records the extent to which a project j depends on another project i.

---

4. This is referred to as a dependency matrix in Reference [7]; however, I decided to call it a propagation matrix in order to avoid any confusion with the dependencay matrix described in Chapter 3.

Table 6.12 Propagation Matrix

i	j					
	Project 1	**Project 2**	**Project 3**	...	**Project $n-1$**	**Project n**
Project 1	0	0.25	0		0.30	0
Project 2	0	0	0.90		0.5	0
Project 3	0.30	0	0		0	0
...				0	0	0
Project $n-1$	0	0	0		0	0.25
Project n	0	0	0		0	0

$$P^{n \times n} = [\hat{pi}_j] = \begin{cases} 0 & \text{if the project } j \text{ does not depend on project } i \text{ or if } j = i \\ 0 < x \leq 1 & \text{if project } j \text{ depends on project } i \text{ in degree } x \\ \sum_{i=1}^{n} p_{ik} \leq 1 & \forall k = 1,2,\ldots,n \\ \sum_{j=1}^{n} p_{kj} \leq 1 & \forall k = 1,2,\ldots,n \end{cases}$$

The exact meaning of p_{ij} depends on what it is we are trying to propagate. For example, if we are trying to propagate effort from one project to another, p_{ij} would represent the percentage increase of work in project j to compensate for what is not going to be done in project i, should this be canceled. If what we are trying to propagate is the loss of revenue in project j resulting from the cancellation of project i, the meaning of p_{ij} is the percentage of the total revenue of project j that would not be realized in the event i is canceled. Each dimension, the consequences of which we would like to propagate through the portfolio, would require a different matrix.

Table 6.13 details the process by which the effects of a decision are propagated across the portfolio.

Although comparing the rows and columns of the propagation matrix could signal how strongly a project influences the outcomes of other projects in the portfolio, or how dependent a project is on other projects, these

Table 6.13 Propagation of Effort

Step	Inputs and Process	Result
1	The additional work to be done in project j should project i be cancelled is estimated and recorded in the propagation matrix P^{nxn} as a fraction of project i effort	(matrix below)
2	Assuming the existence of two vectors E and F $E^n = [e_i]$ = effort required by project i $$F^n = [f_i] = \begin{cases} 0 \text{ if project } i \text{ is not funded} \\ 1 \text{ if project } i \text{ is funded} \end{cases}$$	(vectors below)
3	For each funded project, create an activity-on-vertex network (algorithm not shown) with all non-funded projects that are directly or indirectly connected to the funded project.	(network below)

Step 1 matrix:

	Project 1	Project 2	Project 3	Project 4	Project 5
Project 1		.20		.10	
Project 2			.20	.10	
Project 3					.4
Project 4					
Project 5					

Step 2 vectors:

	Effort in '000 man-hours		Funding Available
Project 1	10	Project 1	0
Project 2	20	Project 2	0
Project 3	30	Project 3	1
Project 4	40	Project 4	1
Project 5	20	Project 5	1

indicators alone must be used with caution, since a large number of small dependencies could be offset by a single dependency that affects a very large project.

Table 6.13 (continued)

Step	Inputs and Process	Result
4	The additional effort to be added to the funded projects is the sum associated with each of the cancellations according to the following rules: j is the project under analysis so for all the immediate nodes i, $Arc_{ij} = E_j \cdot P_{ij}$ i is a predecessor of project j so for all other nodes in the tree $Arc_{ki} = E_i \cdot d_{ik}$ where $d_{ik} = \dfrac{P_{ik}}{\sum\limits_{\alpha=1}^{n} P_{i\alpha} \cdot F_{\alpha}}$ Allocations different from those based on the normalized weights d_{ik} are possible, but this seems to be a fair policy.	
5	The recalculated efforts are as follows	(see table below)
6	Calculate propagation metrics (e'_i) is the recalculated effort for project i in step 5, and e_i is the original effort vector) For $i = s$ to n if $\hat{e}i \neq e'_i$ then $NetSavings = NetSavings + (\hat{e}i - e'_i)$ $DisruptionFanOut = DisruptionFanOut + 1$ endif Next i	$NetSavings = 12.1$ $DisruptionFanOut = 4$

Result for Step 5:

	Original effort in '000 man-hours	Additional effort in '000 man-hours	Re-calculated effort
Project 1	10	Non-funding	0
Project 2	20	Non-funding	0
Project 3	30	8.66	38.66
Project 4	40	9.33	49.33
Project 5	20	0	20

6.3.4   The portfolio approach

The reason for instituting project portfolio management in an organization is to select and execute those projects that balance its conflicting needs, such as high returns and low risk, current versus future returns, capacity to quickly respond to new demands and efficiency, and so on. Consequently, once the benefits to be derived from the execution of individual projects has been evaluated, or revisited in the case of projects already in the portfolio, it is time to decide how the organization's resources should be allocated across the different projects in order to achieve the balance sought.

6.3.5 Building visual maps

Visual mapping tools use graphical and charting techniques to simultaneously portray the benefits and costs of the projects under consideration. Probably the most conspicuous of these is the bubble diagram (see Figure 6.11) that shows the projects in a two-dimensional plot [9], using the size and sometimes the color or shape of the "bubbles" to convey additional project information. Each of the axes of the diagram corresponds to a dimension on which the project is ranked.

Generally, two bars are drawn dividing the chart space into four quadrants. Each bar defines a threshold that projects must exceed in order to be included in the portfolio. The most popular diagram is a risk-return diagram (such as the probability of technical success and reward). In this case, each quadrant represents a different combination of risk/return: low-risk/

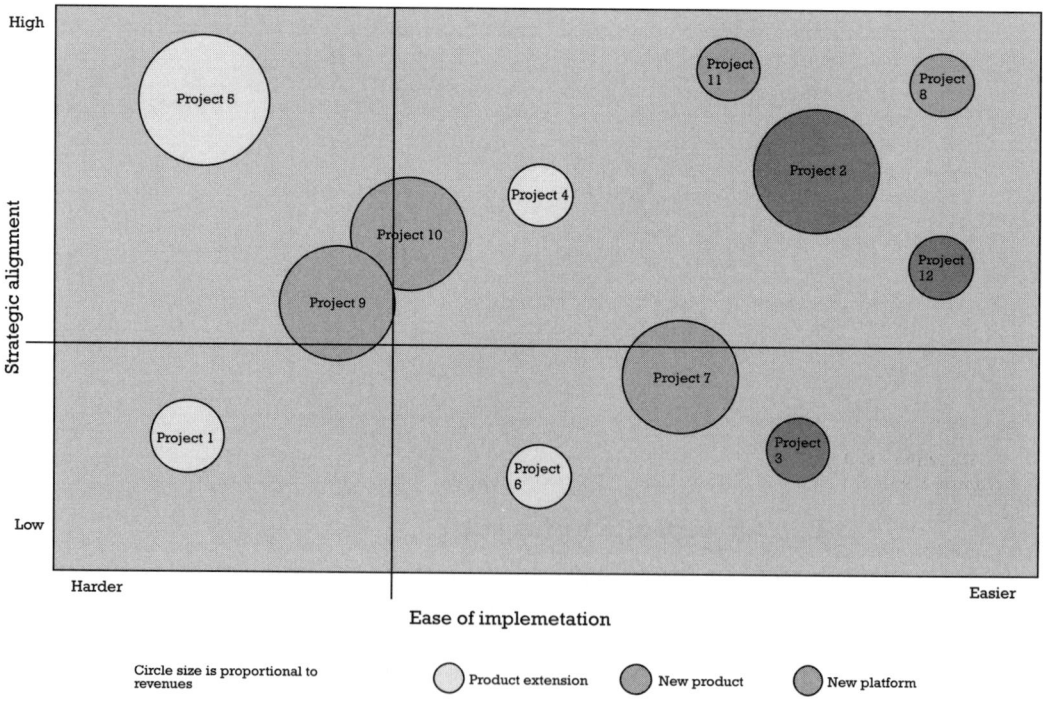

Figure 6.11 Bubble diagram.

low-return; high-risk/high-return; low-risk/high-return; high-risk/low-return.

Other visual mappings (see Figure 6.12) show the number of resources required over time as a function of technology or stage of innovation of the products the projects support.

Because of their ability to portray in a simple way complex relationships between or among projects, visual maps are one of the most popular tools employed in balancing the project portfolio.

6.3.6 Ranking projects

The paired comparisons method (see Figure 6.13) ranks projects against one another in one or more dimensions. In its simplest form the method uses a single rating (i.e., which of two projects being compared is better with respect to a certain attribute); in sophisticated approaches, the method allows for a value specifying how much better (i.e., twice as good, three times better), one project is with respect to the other.

The output of the paired comparisons method is a matrix or table (see Table 6.14) called a judgment matrix, which contains the result of the comparisons between all possible pairs of projects for each dimension being evaluated.

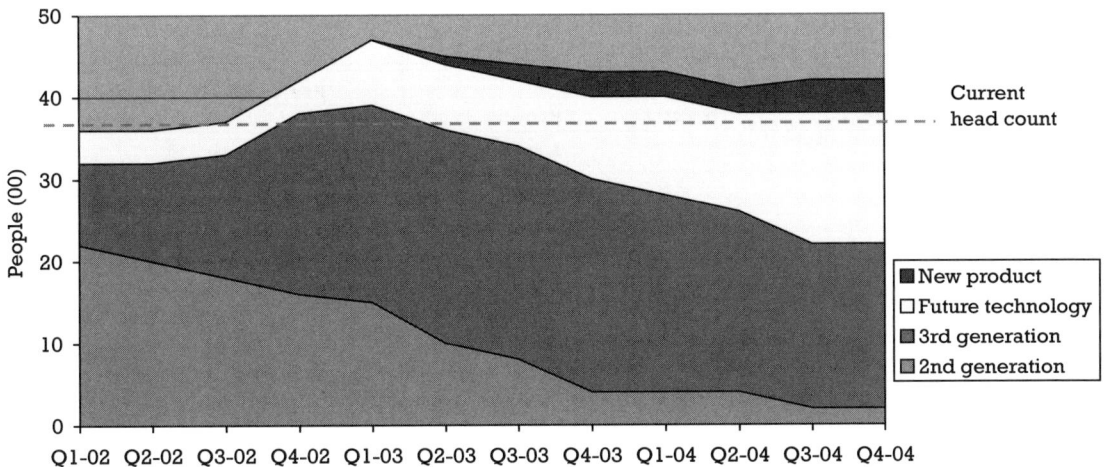

Figure 6.12 Resource-requirements map.

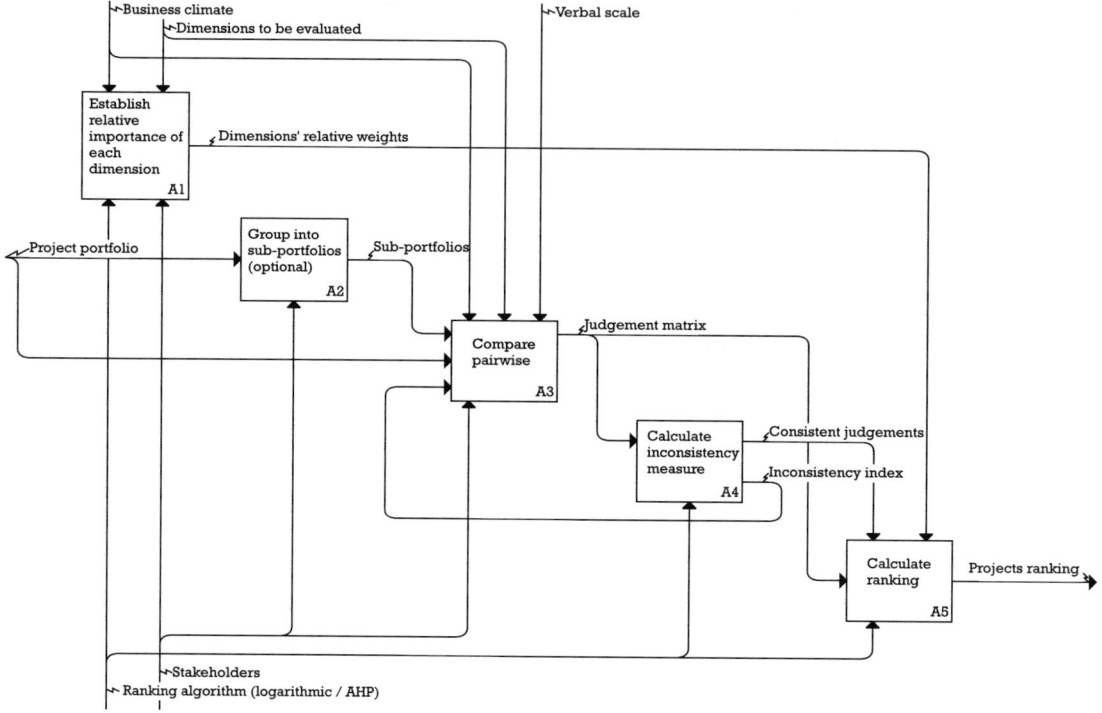

Figure 6.13 Paired comparisons process.

Table 6.14 Judgment Matrix

	Project 1	Project 2	Project 3	...	Project $n-1$	Project n
Project 1	1	1	1.5	...	3	5
Project 2	1	1	1.5	...	2	5
Project 3	0.67	0.67	1	...	2	4
...	...	...	...	1	...	...
Project $n-1$	0.33	0.5	0.5	...	1	1.5
Project n	0.2	0.2	0.25	...	0.67	1

$$A^{nxn} = [a_{ij}] = \begin{cases} a_{ij} = \dfrac{d_i}{d_j} \\ a_{ii} = 1 \\ a_{ji} = \dfrac{1}{a_{ij}} \end{cases}$$

where $\dfrac{d_i}{d_j}$ is the relative importance of project i with respect to project j
along dimension d as judged or perceived by the evaluator. In Table 6.14, Project 1 is judged to be equally as important as Project 2, $a_{12} = 1$, and Project 1 is judged to be 3 times better than Project $n - 1$, $a_{1n-1} = 3$. Notice that in a perfectly consistent judgment matrix all elements satisfy the condition $a_{ij} \times a_{jk} = a_{ik}$, which is clearly not the case in Table 6.14, where Project 1 is judged to be 1.5 times better than Project 3 and 5 times better than Project n, but Project 3 is judged 4 times better than Project n, whereas in a perfectly consistent judgment it should have been said to be $a_{3n} = \dfrac{a_{13}}{a_{1n}} = \dfrac{5}{1.5} = 3.33$ times
better. The good news is that these inconsistencies, if kept within certain limits, are actually good, since nobody knows what the true value is anyway. Remember that the values are judgments, and by definition, judgments are subjective.

If there are n projects being compared among m dimensions, the total number of comparisons to be made is $m \times \left(\dfrac{m-1}{2}\right) + m \times n\left(\dfrac{n-1}{2}\right)$. To prevent this number from growing too large, which is the major drawback of this method, it might be necessary to break down the total portfolio into subportfolios based on customer, product line, or some other relevant criteria.

A verbal scale like the one shown in Table 6.15 can be used to facilitate the judgment process by avoiding lengthy and futile discussions about whether a particular project is twice or twice and one quarter more important than the other, without affecting its output in a significant way.

Once the judgment matrices, one for each dimension, have been completed, the next step is to verify the consistency of the comparisons and derive a ranking from them. This is the point at which the different methods—logarithmic least squares (see Table 6.16) and the eigenvectors method, better known by its commercial name as the analytic hierarchy process (see Table 6.17)—differ.

Table 6.15 Verbal Scale

Definition	Explanation	a_{ij} value if project i is preferred to project j	a_{ij} value if project j is preferred to project i
Of equal value	The two projects are roughly of equal value.	1	1
Slightly more (less) value	Experience and/or judgment recognizes one project as being somehow more (less) valuable than other.	3	.33
Essential or strong (weak) preference	Experience and/or judgment strongly favor one project over other.	5	.2
Very strong (weak) preference	The dominance of one project over other is self-evident. Dominance is demonstrated in practice.	7	.14
Strongest (weakest) preference	The difference between the projects being compared is of an order of magnitude	9	.11
Intermediate values between adjacent scales	When compromise is needed	2, 4, 6, 8	.5, .25, .16, .12

After: [10]

These methods might seem overly complicated for simple one-dimensional comparisons; however, they become invaluable when it is necessary to produce a ranking of projects across a hierarchy of sometimes conflicting attributes. The mathematics in this case are similar to that presented here, with the result of one dimension being scaled by the relative weights of the dimensions.

6.4 Net present value and the gated project approach

NPV is possibly the most widely accepted criterion for project evaluation. A project with a positive NPV increases the wealth of the firm, since the total value generated through the project's lifetime is superior to the cost of financing it. NPV is measured in today's dollars. The equation below summarizes the process as commonly applied.[5]

---

5. The general form of the NPV equation is $NPV = \sum_{t=0}^{N} \dfrac{\left(Benefits_i - Cost_i\right)}{\prod_{j=0}^{t}(1+i_j)}$, which allows for different discount rates

for different periods. The common form of the equation assumes that the discount rate is the same for all periods.

Table 6.16   Logarithmic Least Squares

Step	Inputs and Process	Result
1	The projects are judged on relative merits recorded in judgment matrix A. Assume 4 projects: Project 1, Project 2, Project 3, and Project 4 whose ratings are as shown in the result column. The judges only need to complete the cell a_{ij} corresponding to the upper diagonal matrix.	<table><tr><th></th><th>Project 1</th><th>Project 2</th><th>Project 3</th><th>Project 4</th></tr><tr><td>Project 1</td><td></td><td>4</td><td>6</td><td>7.5</td></tr><tr><td>Project 2</td><td></td><td></td><td>1.5</td><td>2</td></tr><tr><td>Project 3</td><td></td><td></td><td></td><td>2</td></tr><tr><td>Project 4</td><td></td><td></td><td></td><td></td></tr></table>
2	The matrix is completed by applying the expressions below: $$A^{nxn} = [a_{ij}] = \begin{cases} a_{ij} = \dfrac{d_i}{d_j} \\ a_{ii} = 1 \\ a_{ji} = \dfrac{1}{a_{ij}} \end{cases}$$	<table><tr><th></th><th>Project 1</th><th>Project 2</th><th>Project 3</th><th>Project 4</th></tr><tr><td>Project 1</td><td>1</td><td>4</td><td>6</td><td>7.5</td></tr><tr><td>Project 2</td><td>.25</td><td>1</td><td>1.5</td><td>2</td></tr><tr><td>Project 3</td><td>.16</td><td>.7</td><td>1</td><td>2</td></tr><tr><td>Project 4</td><td>.13</td><td>.5</td><td>.5</td><td>1</td></tr></table>
3	The row's geometric means are calculated: $$v_i = \left(\prod_{j=1}^{n} a_{ij} \right)^{\frac{1}{n}}$$	$$v_i = \begin{bmatrix} 3.6 \\ .93 \\ .68 \\ .42 \end{bmatrix}$$
4	The projects ranking is calculated: $$r_i = \frac{v_i}{\sum_{l-1}^{n} v_l}$$ r_i measures the relative importance of each project in relation to the others, the higher the value, the more important the project is: Therefore, the ranking is Project 1, Project 2, Project 3, Project 4	$$\sum v_i = 5.7$$ $$\begin{bmatrix} \dfrac{3.6}{5.7} \\ \dfrac{.93}{5.7} \\ \dfrac{.68}{5.7} \\ \dfrac{.42}{5.7} \end{bmatrix} = \begin{bmatrix} .64 \\ .16 \\ .12 \\ .07 \end{bmatrix}$$

$$NetPresentValue = \sum_{1}^{n} \frac{(Benefits_t - Cost_t)}{(1 + i)^t}$$

Table 6.16 (continued)

Step	Inputs and Process	Result
5	The inconsistency index is calculated: $$InconsistencyIndex = \dfrac{\sqrt{\displaystyle\sum_{i=1}^{n}\sum_{j>1}^{n}\left(\ln a_{ij} - \ln \dfrac{v_i}{v_j}\right)^2}}{\dfrac{(n-1)(n-2)}{2}}$$	$InconsistencyIndex = 3\%$
6	If the consistency index exceeds 10% the offending judgments are found using the expression $a_{ij} \times a_{jk} <\!<\!>\!> a_{ik}$ and reviewed	

Table 6.17 Analytical Hierarchy Process

Step	Inputs and Process	Result
1	The projects are judged on relative merits recorded in judgment matrix A. Assume 4 projects: Project 1, Project 2, Project 3, and Project 4 whose ratings are as shown in the result column. The judges only need to complete the cell a_{ij} corresponding to the upper diagonal matrix.	<table><tr><th></th><th>Project 1</th><th>Project 2</th><th>Project 3</th><th>Project 4</th></tr><tr><td>Project 1</td><td></td><td>4</td><td>6</td><td>7.5</td></tr><tr><td>Project 2</td><td></td><td></td><td>1.5</td><td>2</td></tr><tr><td>Project 3</td><td></td><td></td><td></td><td>2</td></tr><tr><td>Project 4</td><td></td><td></td><td></td><td></td></tr></table>
2	The matrix is completed by applying the expressions below: $$A^{nxn} = [a_{ij}] = \begin{cases} a_{ij} = \dfrac{d_i}{d_j} \\ a_{ii} = 1 \\ a_{ji} = \dfrac{1}{a_{ij}} \end{cases}$$	<table><tr><th></th><th>Project 1</th><th>Project 2</th><th>Project 3</th><th>Project 4</th></tr><tr><td>Project 1</td><td>1</td><td>4</td><td>6</td><td>7.5</td></tr><tr><td>Project 2</td><td>.25</td><td>1</td><td>1.5</td><td>2</td></tr><tr><td>Project 3</td><td>.16</td><td>.7</td><td>1</td><td>2</td></tr><tr><td>Project 4</td><td>.13</td><td>.5</td><td>.5</td><td>1</td></tr></table>

Table 6.17 (continued)

Step	Inputs and Process	Result
3	Normalize the columns of the judgment matrix: $$\begin{bmatrix} \delta_{11} = a_{11} / \sum_{i=1}^{n} a_{i1} & \delta_{1n} = a_{1n} / \sum_{i=1}^{n} a_{in} \\ \dots & \dots \\ \delta_{n1} = a_{n1} / \sum_{i=1}^{n} a_{i1} & \delta_{nn} = a_{nn} / \sum_{i=1}^{n} a_{in} \end{bmatrix}$$	$$\begin{bmatrix} \frac{1}{1.54} & \frac{4}{6.2} & \frac{6}{9} & \frac{7.5}{12.5} \\ \frac{.25}{1.54} & \frac{1}{6.2} & \frac{1.5}{9} & \frac{2}{12.5} \\ \frac{.16}{1.54} & \frac{.7}{6.2} & \frac{1}{9} & \frac{2}{12.5} \\ \frac{.13}{1.54} & \frac{.5}{6.2} & \frac{.5}{9} & \frac{1}{12.5} \end{bmatrix} = \begin{bmatrix} .65 & .65 & .67 & .60 \\ .16 & .16 & .17 & .16 \\ .10 & .11 & .11 & .16 \\ .08 & .08 & .06 & .08 \end{bmatrix}$$
4	Calculate the sum of the rows: $$\begin{bmatrix} \beta_1 = \sum_{j=1}^{n} \delta_{1j} \\ \beta_2 = \sum_{j=1}^{n} \delta_{2j} \\ \dots\dots \\ \dots\dots \\ \beta_n = \sum_{j=1}^{n} \delta_{nj} \end{bmatrix}$$	$$\begin{bmatrix} 2.57 \\ .65 \\ .48 \\ .30 \end{bmatrix}$$
5	Calculate the rankings of the projects: $$r_i = \begin{bmatrix} r_1 = \frac{\beta_1}{n} \\ r_2 = \frac{\beta_2}{n} \\ \dots\dots \\ \dots\dots \\ r_n = \frac{\beta_n}{n} \end{bmatrix}$$ r_i measures the relative importance of each project in relation to the others, the higher the value, the more important the project is: Therefore, the ranking is Project 1, Project 2, Project 3, Project 4	$$\begin{bmatrix} \frac{2.57}{4} \\ \frac{.65}{4} \\ \frac{.48}{4} \\ \frac{.30}{4} \end{bmatrix} = \begin{bmatrix} .64 \\ .16 \\ .12 \\ .08 \end{bmatrix}$$

where

i = discount rate;

t = period in which Benefits$_t$ and Cost$_t$ are incurred.

Table 6.17   (continued)

Step	Inputs and Process	Result
6	Calculate the consistency ratio (CR) for the judgments. The CR is the ratio of consistency index (CI) to the random index (RI) for the same-order matrix. The consistency indices of randomly generated reciprocal matrices from the scale 1 to 9 are called the RIs. The RI for matrices of order n are given below. The first element corresponds to a matrix of order 1, the second to a matrix of order 2 and so on. $$\begin{bmatrix} \kappa_1 \\ \kappa_2 \\ \kappa_n \end{bmatrix} = \begin{bmatrix} a_{11} & & a_{1n} \\ a_{21} & & \\ a_{n1} & & a_{nn} \end{bmatrix}\begin{bmatrix} r_1 \\ r_2 \\ r_n \end{bmatrix}$$ $$\lambda = \frac{\sum_{i=1}^{n} \kappa_i / r_i}{n}$$ $$CI = \frac{\lambda - n}{n-1}$$ $$RI = \begin{bmatrix} 0 & 0 & 0.58 & 0.90 & 1.12 & 1.24 & 1.32 \\ 1.41 & 1.45 & 1.49 & 1.51 & 1.48 & 1.56 & 1.57 & 1.59 \end{bmatrix}$$ $$CR = \frac{CI}{RI}$$ A CR of .10 or less is considered acceptable.	$$\begin{bmatrix} 1 & 4 & 6 & 7.5 \\ .25 & 1 & 1.5 & 2 \\ .16 & .7 & 1 & 2 \\ .13 & .5 & .5 & 1 \end{bmatrix}\begin{bmatrix} .64 \\ .12 \\ .16 \\ .08 \end{bmatrix} =$$ $$\begin{bmatrix} 2.68 \\ .68 \\ .50 \\ .30 \end{bmatrix} \cdot \begin{bmatrix} \frac{1}{.64} \\ \frac{1}{.12} \\ \frac{1}{.16} \\ \frac{1}{.08} \end{bmatrix} = \begin{bmatrix} 4.18 \\ 5.66 \\ 3.16 \\ 3.79 \end{bmatrix}$$ $$\lambda_{max} = \frac{4.18 + 5.66 + 3.16 + 3.79}{4} = 4.20$$ $$CI = \frac{4.20 - 4}{3} = .067$$ $$CR = \frac{.067}{.90} = .074$$

Its computation is based on the principle of discounting: All projected future cash flows of the project are discounted back to the present time under the assumption that one dollar today is worth $(1 + i)^t$ dollars at time t in the future. The cash flows represent the estimated costs, cost savings, and revenues at various points during the useful lifetime of the project. The variable i is the discount rate; it captures the opportunity cost and the risk of the underlying investment. A higher NPV is always preferable to a lower NPV, and a negative NPV represents an unacceptable investment [11].

Applying the NPV method to the project illustrated in Figure 6.14, with a discount rate set at 25%[6] and the most likely return of $7,000,000, yields a valuation of

---

6. Lucent in its financial staements for 2001 used risk-adjusted discount rates ranging from 25% to 35%. Boeing for the same year used a composite rate of 18.7% to discount forecasted revenues.

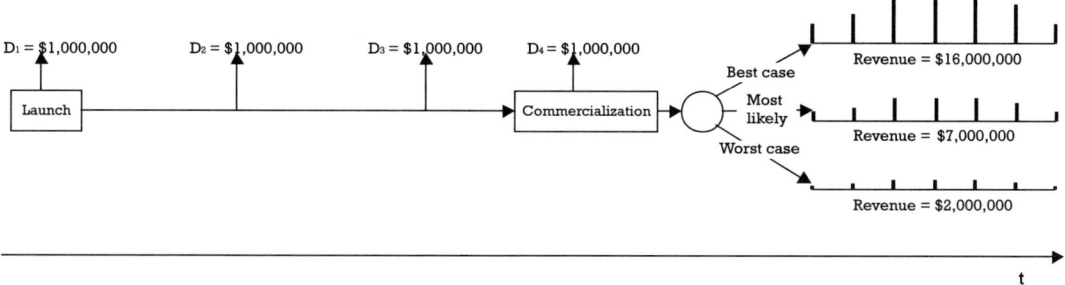

Figure 6.14 NPV project valuation.

$$= -\left(1,000,000 + \frac{1,000,000}{1.25} + \frac{1,000,000}{1.25^2} + \frac{1,000,000}{1.25^3}\right) + \frac{7,000,000}{1.25^4} = -84,8000\$$$

and hence the project is rejected.

The problem with this valuation is that it assumes that the project is executed in its entirety and that the success or failure is only known at the end. This completely ignores the fact that most R&D projects are managed and reviewed throughout their existence and that many of them, but perhaps not enough, are killed in their infancy after changes in market conditions or negative results in the development stage affect their prognosis. Furthermore, in a gated approach to project management the option to continue funding the project, to defer further funding until certain conditions are met, or to stop funding are on the table at each tollgate review. The decision-tree valuation approach (see Figure 6.15) better matches the escalation of commitments implicit in the gated approach.

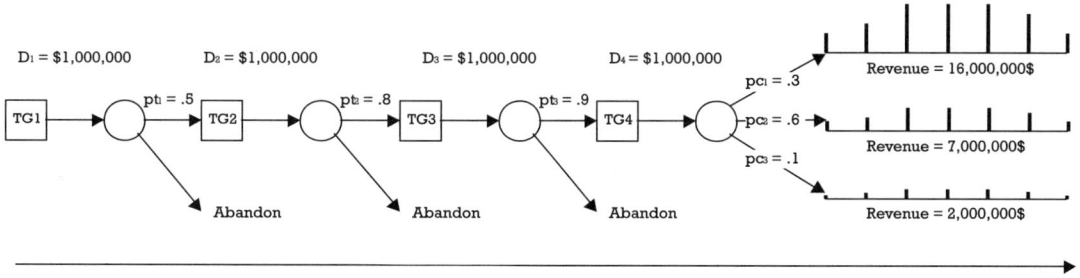

Figure 6.15 Decision-tree project valuation.

In this case, the discount rate would be set at the weighted average cost of capital $(WACC)^7$ of $12\%^8$ and the valuation of this project would be

$$= (E(PV(Revenues)) - PV(D_4))pt_1pt_2pt_3 -$$
$$(D_1 + PV(D_2)pt_1 + PV(D_3)pt_1pt_2)$$

$$E(PV(Revenues)) = 0.3 \times \frac{16,000,000}{(1+0.12)^4} + 0.6 \times \frac{7,000,000}{(1+0.12)^4} +$$

$$0.1 \times \frac{4,000,000}{(1+0.12)^4} = \$5,973,869$$

$$= \left(5,973,869 - \frac{1,000,000}{(1+0.12)^3}\right)0.5 \times 0.8 \times 0.9 -$$

$$\left(1,000,000 + \frac{1,000,000}{(1+0.12)} \times 0.5 + \frac{1,000,000}{(1+0.12)^2} \times 0.5 \times 0.8\right) = \$129,046$$

which provides a return of $3.46\%^9$ over the cost of capital, and hence the project is accepted.

The reason for using different discount rates is that in the NPV method, risk is embodied in the discount rate, while in the decision-tree valuation it is explicitly captured by the probabilities of success attached to each node.

The examples above clearly show how NPV's failure to recognize that management does in fact manage, undervalues projects by not recognizing the reduction in risk that characterizes the gated decision process.

Be aware that the justification for a decision-tree valuation is provided by the ability and willingness of management to kill or change projects that do not perform up to expectations. If for cultural or political reasons management were not prepared to do it, then a decision tree or an options valuation would be inappropriate.

---

7. The after-tax weighted average cost of capital (WACC) is the intyerest rate that must be paid for the investment capital to the stockholders and the debt financiers. It is calculated by the formula *WACC=PecentOfDebtFinancing × CostOfDebtFinancing × (1 − CorporateTaxBracket) + PercentOfEquityFinancing × CostOfEquityFinancing*. The value of WACC is normally provided by corporate finance.

8. The cost of capital for several leading firms for 1999 was AOL, 16.7%; Intel, 12.9%; Pfizer, 11.4%; Walt Disney, 10.0%; and SBC Comm., 8.4%. *Source: Fortune* magazine.

9. Assuming a risk-free interest rate of 5.0%.

6.5   Real options

The term *real option* is used to mean three different things [12]: A calculation method based on or derived from the Black-Scholes[10] equation used in derivatives trading [13], as a means of deriving risk assessments from market-traded securities and commodities [14], and a tool for framing decisions [15].

The original concept of real option valuation is derived from its financial counterpart, where an option represents a right, but not an obligation, to buy or sell something at a predefined price on or before a certain date. In the financial markets, options are contracts to buy or sell some asset, such as a stock, some commodity, or foreign exchange. In real options, the assets are cash flows instead of financial instruments.

For example, a company might have a contract specifying the right to buy 1 million euros at the price of US $1 per euro up to 1 year from now. The company might pay $50,000 now for such a contract. If at any time the price of the euro rises above $1.05 the company could exercise the option and pocket the difference.

The value of the option arises from the risk asymmetry it purports. Notice that in the example above, the losses the company might incur are limited to $50,000, while the gains are theoretically unlimited. Whatever the exchange rate between the euro and the dollar, the company will still pay $1 for each euro, but only if it benefits from it.

Conceptually, a project with a development phase and a commercialization phase could be considered as an option because the cost of the development gives us the opportunity, but not the obligation, to commercialize the results. The connection between project valuation and option pricing is made (see Figure 6.16) by mapping the project characteristics onto option parameters and then using the Black-Scholes formula for calculating the project value.

---

10. Fisher Black, Myron Scholes, and Robert Merton received the 1977 Nobel Prize in economics for their work on the pricing of derivatives. Their theory and the various forms of their formula are widely used in the trading industry

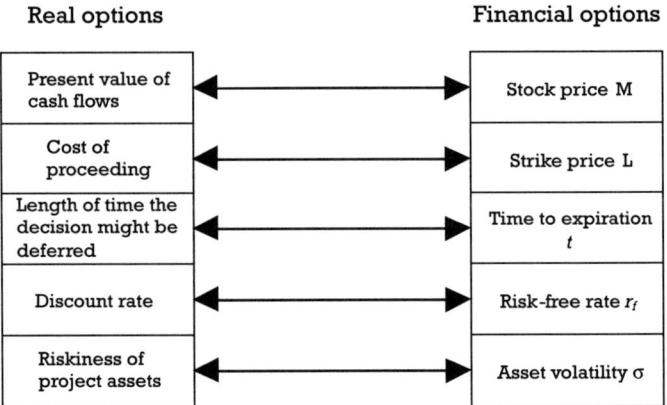

Figure 6.16 Mapping between real and financial options parameters. (*After:* [16].)

$$V = F_N(d_1) \times M - F_N(d_2) \times Le^{(-r_f t)}$$

$$d_1 = \frac{\left(\dfrac{M}{L}\right) + t\left(r_f + \dfrac{\sigma^2}{2}\right)}{\sigma\sqrt{t}}$$

$$d_2 = d_1 - \sigma\sqrt{t}$$

where
V = value of call option on a risk asset = project valuation
L = strike price = cost of proceeding
t = time to expiration = time at which the decision to proceed must be made
r_f = risk-free interest rate = discount rate
M = current price of the asset = present value of project-generated revenues
σ = standard deviation of asset's rate of return = volatility of the generated revenues
$F_N()$ = cumulative probability of a normally distributed variable
e = 2.71

Although the computation of the Black-Scholes formula is straightforward, one must be aware of the assumptions behind it before valuating a project. First, the risk that the formula captures refers to the risk in the returns and not to internal project risks such as staff turnover, lack of funding, or technical difficulties; this still needs to be handled separately [17] with conventional decision trees. The second warning concerns the variability of the generated revenues. In financial options, the variability is obtained from a portfolio of similar assets that are used as a proxy, but in the case of new-product development, such a proxy does not exist or data on it is not readily available. Using the formula outside its domain of application can lead to the making of poor decisions.

As a tool for framing decisions, real options have been used to advocate for building more flexibility into systems [15] and as a justification for not building into systems features that are not needed today [13].

6.6 Summary

Balancing a project portfolio with ten projects that can be arranged in five different ways will result in 120 possible portfolio configurations. This is far beyond what intuition and gut feelings can handle, and this is why we need quantitative techniques to complement business acumen. As with any other stochastic model, the results produced might not always be right, but they will be right most of the time. Or, in the words of Jay Forrester [18], "There seems to be a general misunderstanding to the effect that a mathematical model cannot be undertaken until every constant and functional relationship is known to high accuracy. This often leads to the omission of admittedly highly significant factors (most of the 'intangible' influences on decisions) because these are unmeasured or unmeasurable. To omit such variables is equivalent to saying that they have zero effect. Probably the only value known to be wrong."

References

[1]   Pittiglio, Rabin, and Todd S. McGrath, *A Recipe for Growth in Technology-Based Industries*, PRTM, 1998.

[2]   Pisano, G., *The Development Factory: Unlocking the Potential of Process Innovation*, Cambridge, MA: Harvard Business School, 1996.

[3]   Grey, S., *Practical Risk Assessment for Project Management*, New York: John Wiley & Sons, 1995.

[4]   Goldberg, M., and C. Weber, "Evaluation of the Risk Analysis and Cost Management (RACM) Model," Institute for Defense Analyses, IDA Paper, P-3388, 1998.

[5]   Kindinger, J., "Use of Probabilistic Cost and Schedule Analysis Results for Project Budgeting and Contingency Analysis at Los Alamos National Laboratory," *Proc. 30th Annual Project Management Institute 1999 Seminars and Symposium*.

[6]   Pappas, D., "Contingency Reserves: False Expectations and Misconceptions," Paper presented at the *5th European Project Management Conference*, PMI Europe, Cannes, France, 2002.

[7]   Dickinson, M., et al., "Technology Portfolio Management: Optimizing Interdependent Projects over Multiple Time Periods," *IEEE Trans. on Engineering Management*, Vol. 48, No. 4, Nov. 2001.

[8]   Bussey, L., *The Economic Analysis of Industrial Projects*, Upper Saddle River, NJ: Prentice Hall, 1978

[9]   Cooper, R., *Portfolio Management for New Products*, Cambridge, MA: Perseus Publishing, 2001.

[10] Saaty, T., *The Analytic Hierarchy Process: Planning, Priority Setting, Resource Allocation, Second Edition*, Pittsburgh, PA: RWS Publications, 1996.

[11] Erdogmus, H., and J. Vandergraaf, "Quantitative Approaches for Assessing the Value of COTS-centric Development," Institute for Information Technology, Software Engineering Group National Research Council of Canada.

[12] Lang, J., "Real Options Results Are In: Executives, Beware of the Hype," Executive Briefing, Strategic Decision Group, 2001.

[13] Erdogmus, H., and J. Favaro, *Keep Your Options Open: Extreme Programming and the Economics of Flexibility*, Boston, MA: Addison-Wesley, 2002.

[14] Dixit, A., and R. Pindyck, *Investment Under Uncertainty*, Princeton, NJ: Princeton University Press, 1994.

[15] De Neufville, R., "Real Options: Dealing with Uncertainty in Systems Planning and Design," *5th Int. Conf. on Technology Policy and Innovation*, Technical University of Delft, Netherlands, 2001.

[16] Luehrman, T., "Investment Opportunities as Real Options," *Harvard Business Review*, July-August 1998.

[17] Neely, J., III, "Improving the Valuation of Research and Development: A Composite of Real Options, Decision Analysis and Benefit Evaluation Frameworks," Ph.D. diss., MIT, 1998.

[18] Forrester, J. W., *Industrial Dynamics*, Cambridge, MA: MIT Press, 1961.

Quantitative management

Imagine that you are the PO manager, and that you've just asked a project manager if he or she is going to meet a deadline set for next week. Which of the following answers will provide you with the best information in order to make a decision as to whether something needs to be done?

- "We are a little bit off, but we are going to make it. Don't worry."
- "By this time we had planned to execute 200 test cases of a total of 300, but we were only able to execute 180, so we are a little bit off. However, with 1 week left to go, I think we can recover. Don't worry."

I don't know about you, but to me, the first response would make me more concerned than I'd been before asking the question. The second response, on the other hand, not only provides the information needed to appreciate the current situation but it does so based on an agreed scale known to both the sender and the receiver of the message, which provides the basis for a more objective communication, less prone to errors and misunderstandings. Furthermore, we can manipulate the information received to create new knowledge, such as how long it takes on the average to run a test case, or to forecast a tentative date of completion from the progress so far.

Although a very important aspect of quantitative management, measuring alone is hardly enough. In order to make sense of what we are measuring we need to attach meaning to it. In his book *Quality Software Management,* Gerald Weinberg [1] gives the example of an unusual noise coming from the engine compartment of a car while it is being driven. Such information is accurate, timely, and objective but what does it indicate? Weinberg proposes three possibilities:

1. The driver doesn't hear the noise because he or she is distracted, listening to or thinking about other things.

2. The driver perceives the noise as ominous, but a mechanic would know that it is just the washer fluid vessel that needs to be tightened.

3. The driver perceives the noise as irrelevant, but a mechanic would know that the car is about to run out of oil, damaging the engine.

The driver and the mechanic may have the same information, but the mechanic, based on his knowledge of how engines work, will know what to do with it. He knows what meaning to attach to the noise. Without the meaning, a person can't know what the appropriate response should be.

Projects have a way of making noise. They produce a considerable amount of data, such as the number of hours spent on a task, the number of errors found during testing, the amount of time spent in rework activities, the amount of overtime, and the number of people that requested to be transferred out of the project. Such "noise" can give one insight into what is really going on. But just as in the case of the car, hearing the noise is not enough. To manage based on metrics, we need to supplement the measurements with models that allow us to understand how the project behaves as a system.

As shown in Figure 7.1, forecasting task outcomes and steering the project is just one of the ways we can utilize information collected through measurements. Other uses include employing historical data to estimate and plan new projects, correlating two sets of measurements to understand how processes interact with each other, and producing descriptive and inferential statistics to compare the capabilities of one organization with the capabilities of others for process improvement purposes.

Although we mention some specific metrics and describe them, the purpose of this chapter is not to enumerate everything that could possibly be measured in a project, but to lay the groundwork necessary to establish a fact-based management PO.

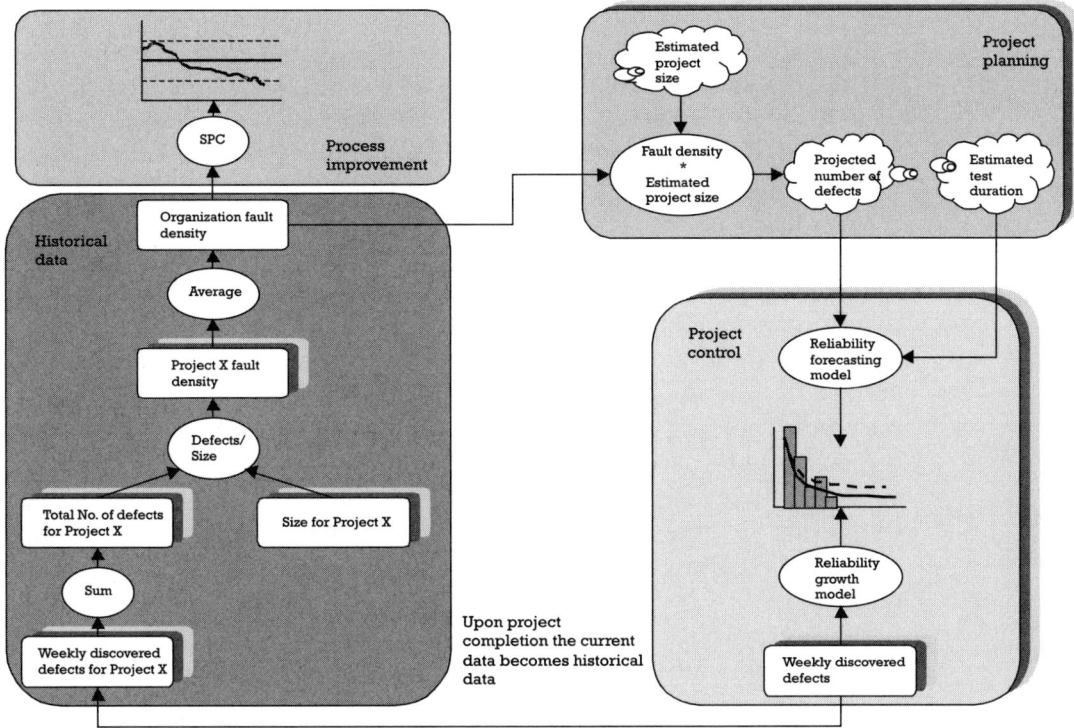

Figure 7.1 Measurements and the multiple purposes they serve.

7.1 Measurement fundamentals

Measuring is not easy. First is the question of the validity and reliability of the measurements. Second is the dilemma of which scale to choose to represent the data captured. Third is the problem of relating these measurements to other measurements. Fourth is the difficulty of aggregating or summarizing the measurements to describe and make inferences about the processes' capabilities. Fifth is the fact that by defining measurements, we are defining what we regard as important, otherwise why would we measure it? And this simple implication might distort the measurements we collect, due to the general tendency of people to put themselves and their projects in the best possible light.

7.1.1 Validity and reliability

For measurements to be useful, they need to be valid and reliable. A valid measurement is one that actually measures what it claims to measure. An example of a common invalid measurement is the number of hours spent as a proxy for progress. A reliable measurement is one that will give you or anyone else approximately the same value when taken on the same object or individual. The commonly used "percentage complete" measurement is highly unreliable because it depends on the opinion of an individual regarding how much work was done and the relative effort required by whatever work is left.

The type of validity to which we referred above is called construct validity and it is of course fundamental to any measurement activity, but the fact that the metric reflects the concept that we want to measure alone does not make it a useful metric. There are other types of validity that need to be verified in order to select metrics that can be relied on for decision-making purposes. These validity types are predictive validity, which is the ability of the metric to be used for estimating and forecasting purposes, discriminant validity, which refers to the ability of the metric to distinguish between things that are different, and content validity, which refers to the extent to which the measure covers all the meanings included in the attribute being measured.

Reliability refers to the consistency of a number of measurements taken using the same measurement method. If the measurements yield approximately the same value, the measurement is reliable. If the variations among them are large, the reliability is low. The reliability of a measurement is influenced by the quality of its definition; vague definitions are likely to result in unreliable measurements, by the measurement instrument and even by the reporting routines.

7.1.2 Levels of measurement

The data used to manage projects or to improve processes is the result of a measurement process that maps the attributes of a task, deliverable, or other relevant entities into a well-defined scale.

The reason for bringing up the topic of scales types, also called measurement levels since the information content associated with each of them differs, is twofold. First, relevant relationships that might exist between objects in the "real world" could be lost in the process of measurement if the scale selected does not possess certain properties. Second, depending on the type of scale selected it would be possible to apply certain transformations like adding, subtracting, or averaging to the measured values and to infer a number of conclusions but not others. In summary, the choice of scale limits the type of information that we can extract from the data collected.

Scales are classified according to whether or not they have the properties of magnitude, equal intervals, and absolute zero [2]. When a scale has magnitude, one instance of the attribute being measured can be judged greater than, less than, or equal to another instance of the attribute. When a scale possesses the property of equal intervals, the distance between consecutive values of the attribute is the same regardless of where in the scale they fall. An absolute zero is a value that indicates that nothing of the attribute being measured exists.

Based on these properties, four types of measurement scales or measurement levels are commonly distinguished: nominal, ordinal, interval, and ratio.

In a nominal scale, the measurement values are categories(i.e., they do not have magnitude). For example, the classification of defects in a software project according to their source does not imply an order among them (i.e., it does not make sense to say that the category "requirements" is less than the category "Coding," this even if the categories were labeled "1" and "2," respectively). When the results of a measurement are expressed in a nominal scale, the type of analysis and summarization that we can do is reduced to counting and establishing proportions among the categories. For a nominal scale, the only measure of central tendency that makes sense is the mode, that is the category with the most occurrences.

Next in the hierarchy of measurements are the ordinal scales. Next means that all operations and transformations applicable to a nominal scale are also applicable to an ordinal scale. An ordinal scale has the property of magnitude, so it is possible to rank objects and arrange them in ascending (or descending) order; however, since an ordinal scale has neither the equal intervals nor the absolute-zero properties, the distances between the values have no meaning. The classification of change requests according to priorities such as "1," "2," and "3," where the ones labeled "1" are more urgent than those labeled "2" and those labeled "2" are more important than those labeled "3," is an example of an ordinal scale. Statements, such as "The average priority for these changes is 2.3" or "a priority 1 request is twice as important as a priority 2 request," however, are inconsistent with the use of an ordinal scale. For ordinal scales, we can use the mode or the median as a measure of central tendency and percentiles as a measure of dispersion.

Next come the interval and the ratio scales. These two scales possess both magnitude and equal intervals. The difference between them is that the ratio scale has an absolute-zero point and the interval scale does not. The classic examples of interval scales are the Celsius and Fahrenheit scales, in which a temperature of 0 degrees does not mean that there is no temperature at all, only that the temperature at that point is colder than a temperature of 10 degrees by a difference of 10 degrees. So in the case of an interval scale, saying that a temperature of 20 degrees is twice as hot as a temperature of 10 degrees would be totally misleading. Imagine that we equate the concept of "warmth" with values on the Celsius scale, and suppose that one Monday we measure a temperature of 10 degrees. If the temperature drops to 1 degree Celsius on Tuesday, was it really ten times as warm on Monday? What if the temperature drops to 0.1 degree Celsius on Wednesday? Was it 10 times as warm on Tuesday, and 100 times as warm on Monday? Clock times, calendar dates, and normalized intelligence scores are examples of frequently used measures from interval scales. The arithmetic mean is the typical measure of central tendency and the standard deviation the measure of the dispersion.

With a true origin or absolute-zero point, division and multiplication become meaningful in the case of ratio scales and all the mathematical operations that we customarily use for real numbers are legitimate. Cost, schedule length, and time between failures are examples of ratio scales. In addition to the operations valid for the interval scales, ratio scales allow for geometric mean, harmonic mean, and percent variation.

7.1.3   Measures of dispersion as an expression of risk

Should we decide to measure productivity, or for that matter any other attribute, in a number of projects we would find that seldom would two of them yield the same number. The reason for this is that behind a simple ratio between the output and the input, there are hidden many circumstances such as people abilities and motivation, task difficulty, undocumented interruptions, scope changes, and external factors that influence the measurement and that cannot be accounted for or separated from the measurements themselves. The more of these special circumstances there are and the greater their influence, the greater the difference will be among measured values (see Figure 7.2).

When this data is used to compute project durations or the effort needed, this variability will be passed on to the plans, and obviously, the higher the variability the higher the risk. Similarly, the assumptions we made about the project being planned will result in added uncertainty.

In interval and ratio scales, this variability is quantitatively expressed by the standard deviation of the set of values, and in consequence the larger the standard deviation, the higher the risk. Other measures of dispersion, such as the range and the percentiles of a distribution, will, although less effectively, also express the degree of uncertainty associated with a set of measurements.

Project	Productivity	Reason
Project A	80	Vendor delivered late
Project B	100	Great team!
Project C	95	It went OK
Project D	90	Nothing unusual
Project E	120	Breakthrough!
Project F	96	Routine work
Project G	97	Good people
Project H	80	Key personnel left
.	.	.
.	.	.
.	.	.
Project X	85	Too many changes
Project Y	100	Saved time by reusing design
Project Z	70	Everything went wrong

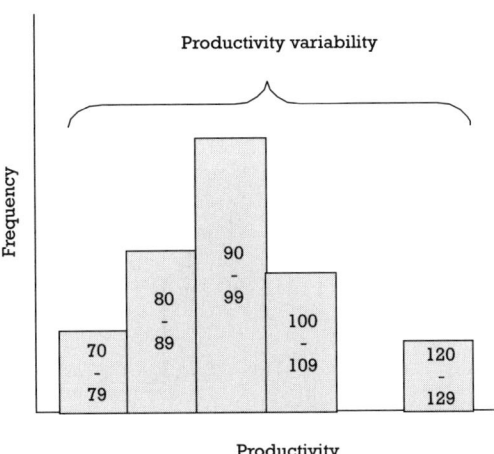

Figure 7.2 Variability present in measurements due to uncontrollable factors.

7.1.4 Relationships between measurement variables

After you have been measuring for a while, certain patterns will start to emerge. Some conditions always seem to lead to the same results, and although they might fail from time to time, they become rather predictable. You might even be tempted to postulate a few theories of your own about how things work or to conjecture the existence of some relationships between a pair of measurements and use them to make decisions.

A scatter plot (see Figure 7.3) is a useful tool to reveal a relationship or association between two measurement variables. Such relationships manifest themselves by any nonrandom structure in the plot.

Scatter plots can provide answers to the following questions:

▸ Are variables X and Y related?

▸ Are variables X and Y linearly related?

▸ Are variables X and Y nonlinearly related?

▸ Does the variation in Y change depending on X?

▸ Are there outliers?

Various common types of patterns are demonstrated by the examples in Figure 7.4.

When there is a relationship between two measurement variables, the variables are called correlated; when there is no relationship, they are called

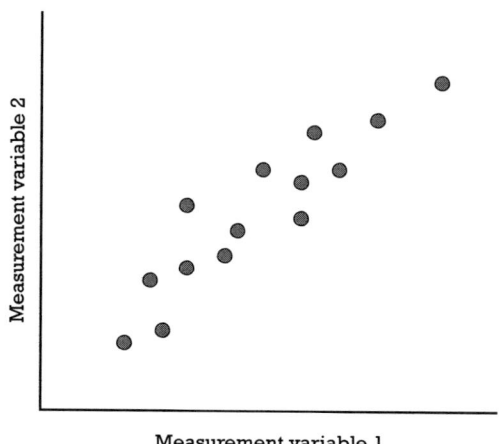

Figure 7.3 Scatter plot showing the relationship between two measurement values.

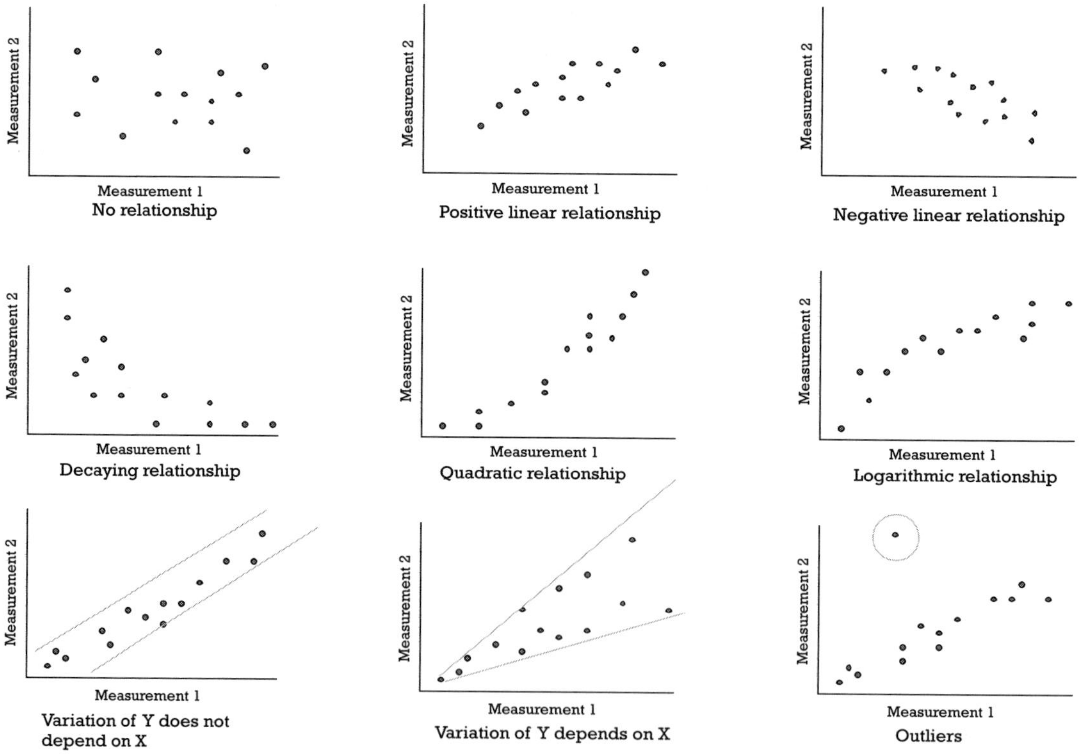

Measurement 2 Measurement 1 No relationship	Measurement 2 Measurement 1 Positive linear relationship	Measurement 2 Measurement 1 Negative linear relationship
Measurement 2 Measurement 1 Decaying relationship	Measurement 2 Measurement 1 Quadratic relationship	Measurement 2 Measurement 1 Logarithmic relationship
Measurement 2 Measurement 1 Variation of Y does not depend on X	Measurement 2 Measurement 1 Variation of Y depends on X	Measurement 2 Measurement 1 Outliers

Figure 7.4 Relationships between measurement variables: typical patterns.

independent. The fact that two variables are correlated does not imply that they are causally connected; sometimes the two variables could be connected through a third one that makes them move in the same direction at the same time. Take the case of effort and schedule in projects—are they correlated? Or is it scope that drives both cost and schedule? Should we fix the scope, what would be more expensive: a project with a compressed schedule or one with a normal schedule? In other cases there is a third variable, called a confounding variable, whose effects on the response variable cannot be separated from those of the explanatory variable. For example, in a recent study [3], only four out of 24 commonly used object-oriented metrics were actually useful in predicting the quality of a software module when the effect of the module size was accounted for. Yet another common problem could be the presence of outliers in the data, which tends to inflate the strength of a relationship. These problems are illustrated in Figure 7.5.

7.1.5 Aggregating measurements

Measurement data is usually generated at relatively low levels of detail within projects. For example, worked hours are usually collected at the work package or task level; similarly, weight and power consumption are properties measured at the module or assembly level, so in order to create a consolidated picture of the whole project or product for analysis and reporting purposes, it is necessary to aggregate or summarize the primitive measurements across different aggregation structures (see Figure 7.6).

The multiproject environment requires the use of one or more of the following aggregation structures, either in their pure form or in combinations with each other, for summarization purposes:

▸ *Portfolio:* Data is aggregated across projects into a single element, which represents the totality of the projects, or into a hierarchy of intermediate elements representing each of the particular subsets of the total portfolio. In an R&D organization subsets of the portfolio could be established based on product lines, in technology, in risk or according to project categories such as technology development, platform development, new application development, application extension, fixes, and so on. This type of aggregation is useful to balance the project mix.

▸ *Organization:* Data is aggregated across the organizational hierarchy mainly for responsibility accounting and resource planning purposes. The structure of this hierarchy will resemble the structure of cost centers and available capacity will be maintained for each of them.

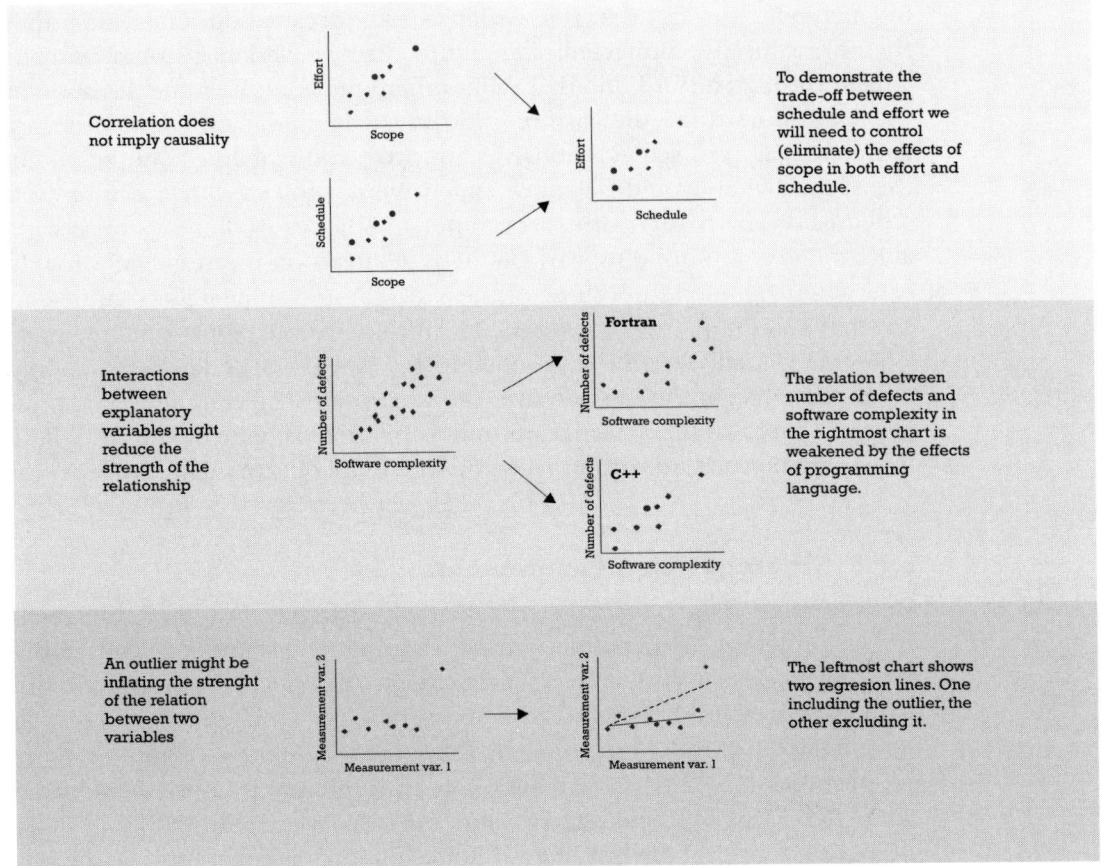

Figure 7.5   Common problems in the analysis of relationships.

- *Project:* Data is aggregated across project activities and or deliverables to provide a consolidated view of what is going on in the project. This is the type of information typically contained in a progress report.

- *Deliverables:* These structures are derived from the relationship of the system components within a particular architecture or design. This structure and the project structure are usually integrated through the project WBS.

- *Activity:* These structures are based on a hierarchy of standardized life-cycle activities that cover the complete activity structure for a project and include tasks such as requirements analysis, design, implementation, integration, and test. This type of aggregation supports activity-based costing (ABC) and activity-based management

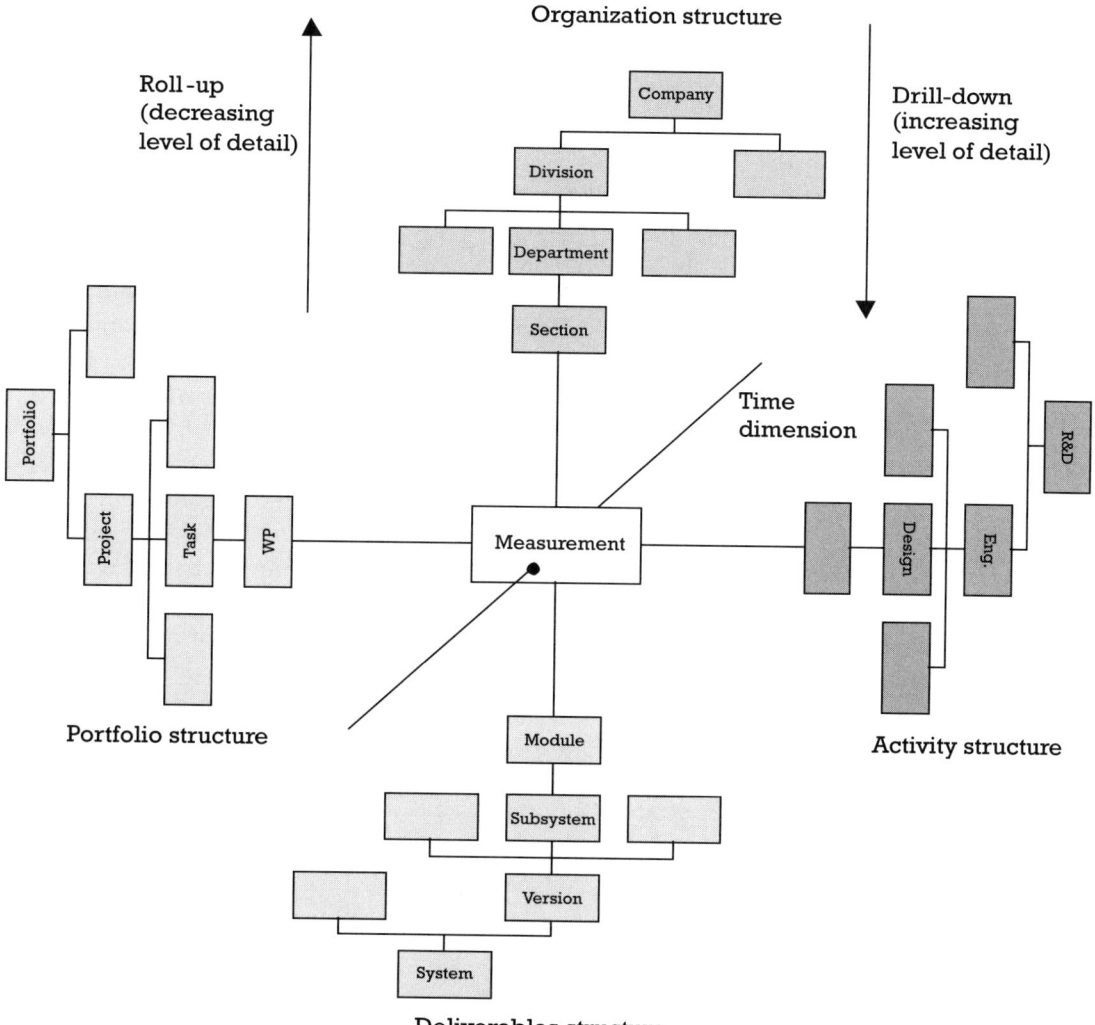

Figure 7.6 Aggregation patterns: Measurements including time report must be available at different levels of aggregation and along the time dimension.

(ABM) and is critical in capturing historical data for estimation and process improvement purposes.

Collectively, the aggregation structures should cover the full spectrum of projects, activities, and deliveries so that every elemental measurement can find its place in the structure, and the intersection between any two of their subsets should be empty to prevent double counting of the same value.

When using aggregated data, there is always the risk that negative variances in one element of the hierarchy could be offset by positive ones in another element, and with everything looking right at the aggregation point, we could be disregarding valuable information about what is happening deep down into the hierarchy. To prevent this, the aggregation processes need to include at least two controls: a measure of the dispersion of the underlying values and a threshold associated with the variances allowable for each element in the aggregation structure.

As the standard deviation of a variable could be used as a quantification of the risk or uncertainty associated with a given variable, it is important to understand what happens to it through the aggregation process. The first thing that needs to be looked at when adding variables is whether there is any relationship between them or if they are independent from a statistical point of view. This is important for two reasons: First, the sum of independent variables yields a much lower risk than the sum of related variables.

In Figure 7.7 we can see that the standard deviation of the sum of 25 fully correlated variables is five times larger than the standard deviation of the sum of the same variables under the assumption that they are independent. The second reason it is important is that the shape of the distribution

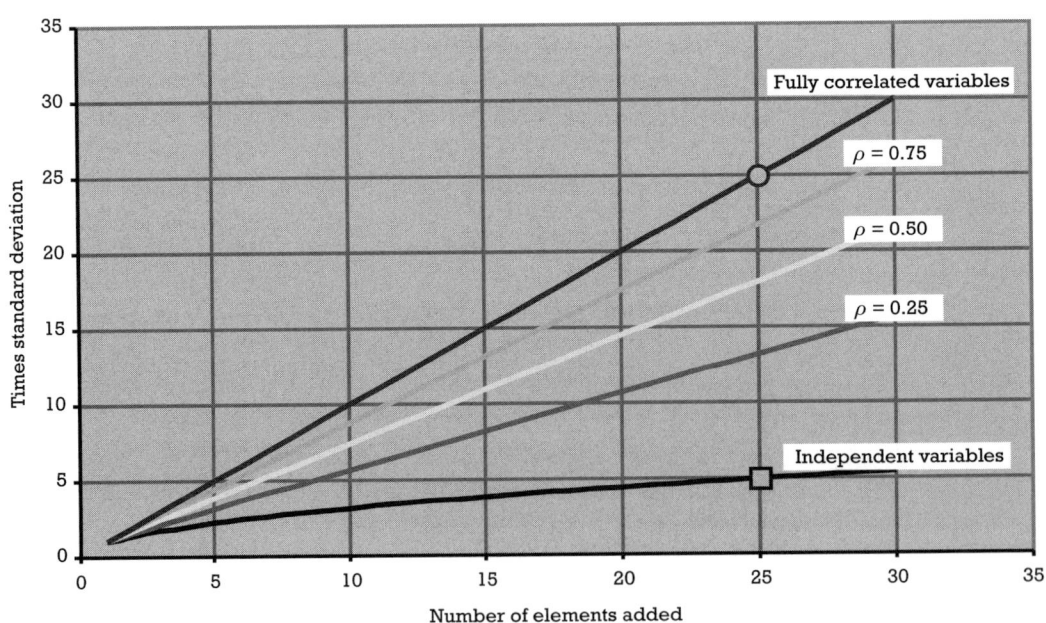

Figure 7.7 Relationship between the number of elements summed, the coefficient of correlation, and the standard deviation of the sum.

will be affected as well. If the variables being aggregated are independent, the shape of the resulting distribution will approach that of a bell-shaped distribution regardless of the shape of the individual distributions. If the variables are correlated, the shape of the sum will instead depend on the shape of the individual distributions and on the strength and nature of their relationship. The reason this happens is simply that when variables are correlated their values tend to move in the same direction at the same time, while in the case of independent variables, the result of this independence is that some values will go up while others go down, and they cancel each other. In practical terms, the assumption of independence is expressed in the belief that the lateness of some tasks is compensated for by the early completion of others and that in the end everything balances out.

This might seem like an academic discussion for some, but it is not difficult to encounter correlated variables in development projects. For example, the underestimation of the system's complexity or the overestimation of the development team productivity will affect the duration of most tasks in the same direction. Thus, if you can think of an underlying cause capable of swinging the measurement values in the same direction, the variables are not independent but correlated, and ignoring this is perhaps one of the most costly mistakes a project manager can make.

7.1.6　Time series

A time series is a chronological sequence of measurements or observations. In a project environment, time series are very important because there is as much information contained in the timing of the measurement as there is in its magnitude. As an example, suppose that, as shown in Figure 7.8, while testing a software system we observe the same number of errors twice, one at the beginning of testing and the other near its end. At the beginning of testing one would expect the results of the next measurement to be larger as we climb through the learning curve and modules are being released for testing. For the second observation, one would expect the following readout to be smaller, as there are no new modules coming in and it is becoming increasingly difficult to find new errors.

The information contained in the time dimension of the measurements is key to the understanding of the process underlying the observed data and this knowledge is essential to the creation of forecasting models. As an example of the practical application of the knowledge gained through the study of time series, take a look at the production data in Figure 7.9. In the three illustrations contained in the picture, we can see that progress does not occur at a constant rate and that it more closely resembles the shape of

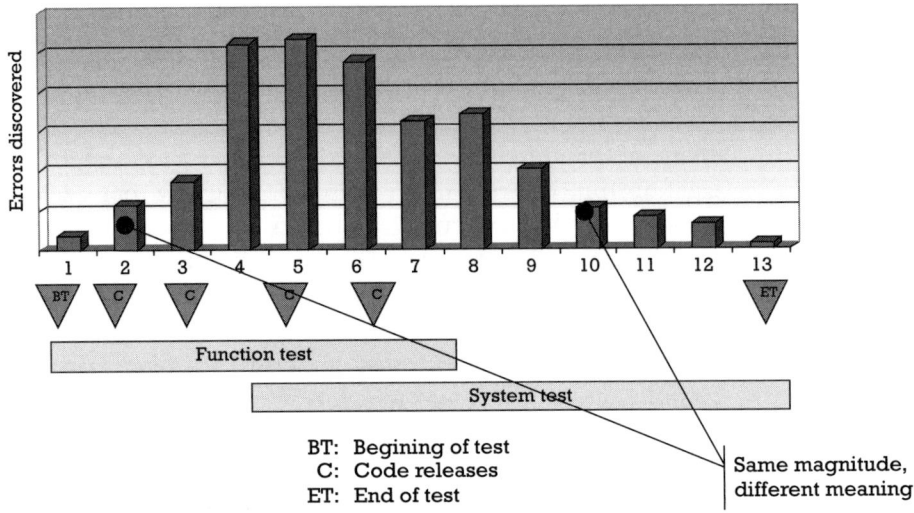

Figure 7.8 Timing of the measurement contains as much information as its magnitude.

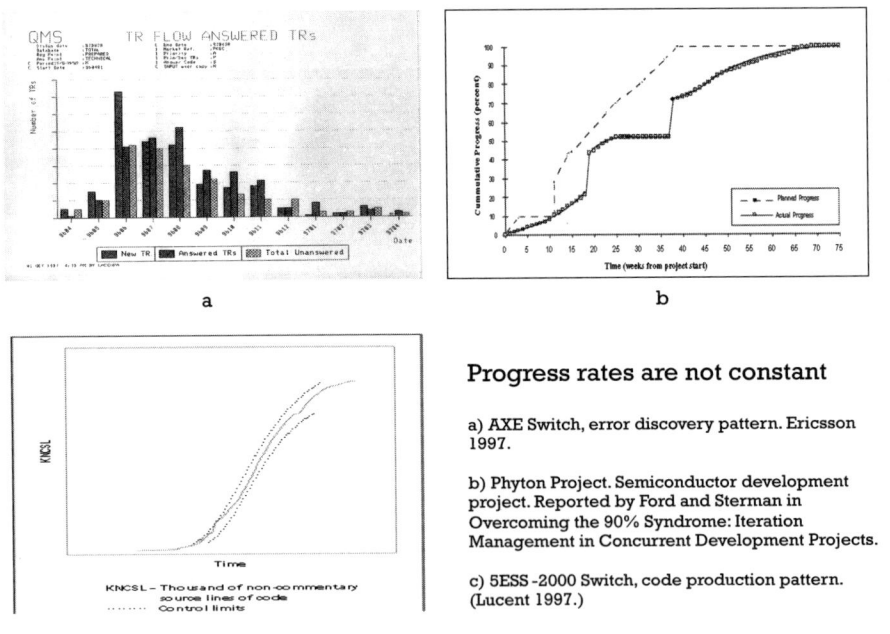

Progress rates are not constant

a) AXE Switch, error discovery pattern. Ericsson 1997.

b) Phyton Project. Semiconductor development project. Reported by Ford and Sterman in Overcoming the 90% Syndrome: Iteration Management in Concurrent Development Projects.

c) 5ESS-2000 Switch, code production pattern. (Lucent 1997.)

Figure 7.9 Progress, measured in terms of its visible output, is not constant through the duration of a task or project.

Figure 7.10. This "S" pattern, typical of many intellectual activities, can be explained by the existence of a number of actions and thought processes at the beginning and end of the task which, although value adding, do not contribute directly to the output being measured, be it number of newly detected problems, thousands of lines of code, or pages written. This knowledge in turn can be used to forecast the completion date of a task more accurately than a linear extrapolation derived from the rate of progress observed through the half-life of the task. In Figure 7.10, production does not grow at a constant rate. At the peak of productivity, between weeks 3 and 5, the percentage complete soars 20% in just 1 week. Toward the end of the task it takes triple the time to go from 80% to 100% complete. Figure 7.11 shows the error incurred by using a linear forecast instead of the S-curve paradigm. Assuming that the task output is 250 units of production (requirements, FP, errors detected, etc.), a linear projection would forecast its completion by week 7.5 while the S curve would put it at week 9. Assuming that the task duration was originally estimated to be 7 weeks, according to the linear projection it will be completed on time, but according to the S curve it will be 2 weeks late.

7.1.7   Sources of data

Obviously, the source of the measurement data will depend on what is being measured; however, it is possible to identify a number of important

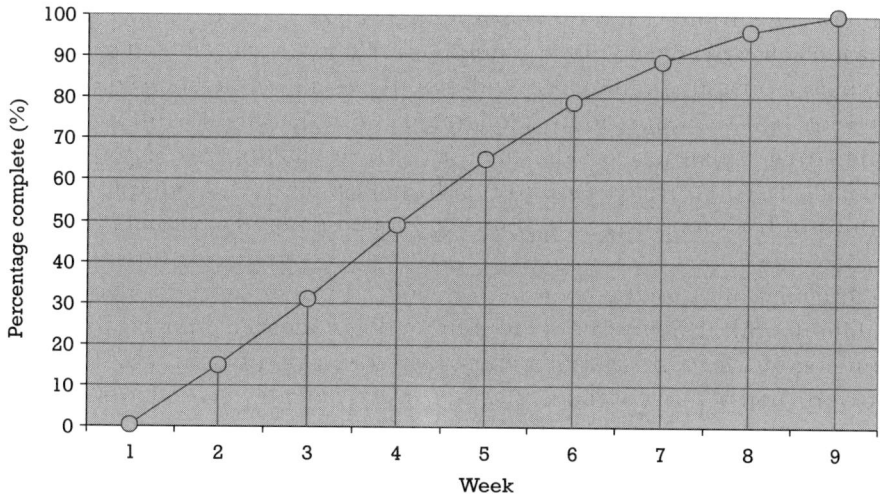

Figure 7.10   The S curve.

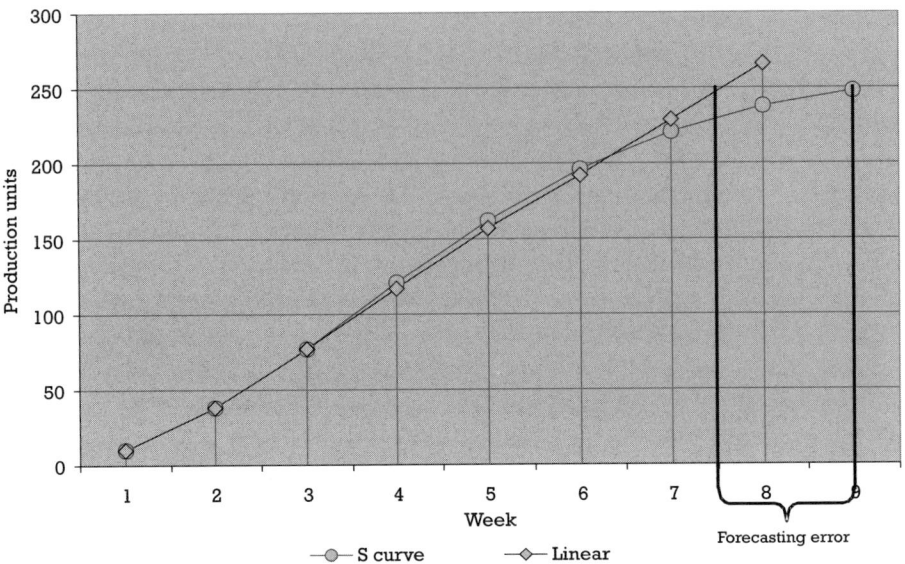

Figure 7.11   Error incurred by using linear forecast instead of S curve.

sources of data in a project environment. These are the time-reporting system, the version control or document management system, and the trouble-
or defect-reporting system. The importance of these sources resides not only
in the wealth of information they can provide, but also in the fact that they
are readily available in most organizations.

By appropriately codifying the hours reported, the time reported could
be used to measure cost, staffing rates, cost of quality, staff disposition, and
organization sustainability. Similarly, the version control or document management system can be supplemented with "probes" or scripts to measure
changes in documents or code every time an artifact is checked in or out.
The trouble-reporting system is also an extremely valuable source of information, not only with respect to the number of defects reported, but in
terms of the rates at which problems are discovered and fixed.

7.1.8   Intrusive nature of measurement

Measuring performance does in fact influence performance. Peter Drucker
[4] has stated that performance measurement in a social system is neither
objective nor neutral. Implicitly, performance measures are a reflection of
what the organization considers important; if quality is not measured, quality must not be important. This can lead to the unintended consequence of

maximizing certain parameters of the organization at the expense of other equally important but much less visible parameters. Examples of unintended consequences include cutting of quality activities to meet a deadline, knowing that the work will have to be redone, at least partially, after the deadline is passed; running shorter test sequences to show a diminishing number of new trouble reports (TRs); and recording expenses as capital expenditures.

Another risk is that when organizations put too much emphasis on measurement for performance purposes, maximizing the measurements might become a substitute for achieving the goals. The way to prevent these unintended consequences is by explaining the purpose of measuring and by not having immediate payoffs linked to achieving certain performance targets and by avoiding as much as possible the use of proxy or indirect measurements instead of direct ones.

7.2 Using metrics

The following sections describe the different uses we can give to metrics in a project environment.

7.2.1 Controlling and steering projects

In this case, measurements are used to ascertain where the project is in relation to the original plan, and to decide what actions, if any, to take to correct the course or to profit from an unplanned advantageous position.

Controlling and steering is performed by monitoring those project dimensions (i.e., progress, performance, people, product, and customer satisfaction) that the stakeholders regard as critical to the success of the project. Tight project budgets and schedules leave little room for recovery. Once things go awry, it is almost too late to fix them within the original project constraints. Effective and efficient steering requires mechanisms, such as those proposed in Section 7.1.6, that can help identify early signs of trouble before it is too late.

7.2.2 Estimating

A historical database of how much effort and how long it took to perform similar activities in past projects could be of invaluable help in planning new projects. Historical data could be used in the construction of parametric estimation models. These models could be later used to forecast effort and lead time as a function of one or more variables such as size and complexity. Estimation models, however, do not have to be limited to the forecasting of effort and duration; they could be used to estimate the number of errors to be corrected, the weight, the power consumption, or any other technical parameter relevant to the project.

Although the specific steps and the methods used to create parametric estimation models depend on the objectives of the model itself, most are built following a process similar to the one depicted in Figure 7.12.

After the data is collected, some kind of exploratory data analysis is conducted to find the activities' cost drivers; the data is then regressed with

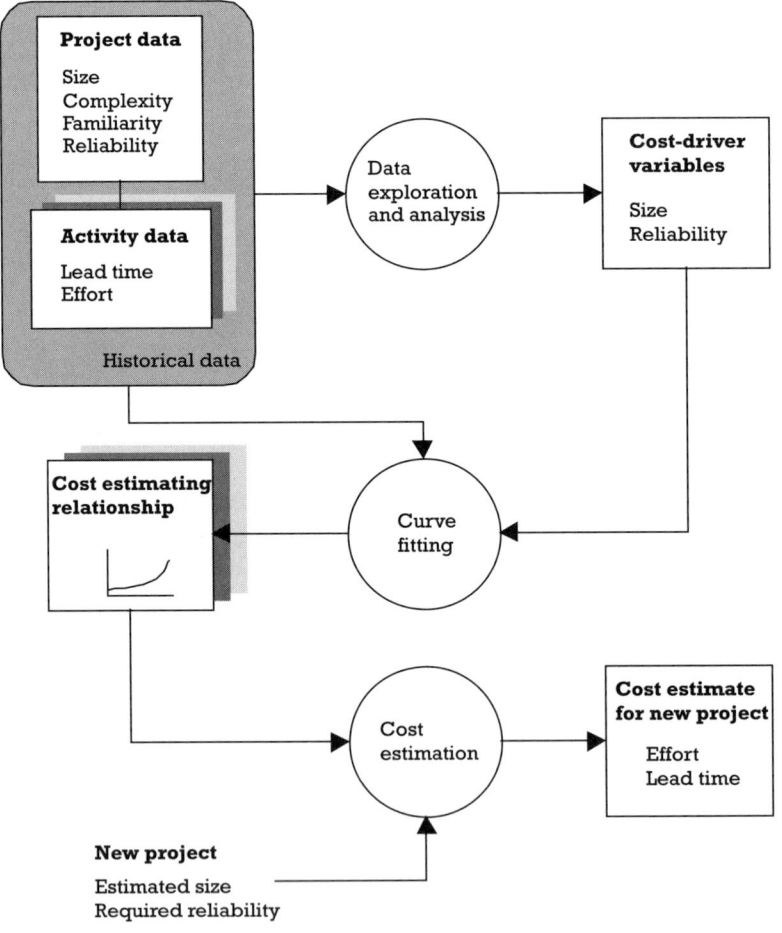

Figure 7.12   Using historical data to build an estimation model.

respect to these drivers and a set of equations called cost estimation relation-ships is derived. The equations are later used to estimate new projects.

When collecting data for estimation purposes it is necessary to capture, together with the measurements of the attributes we are interested in, the context information necessary to normalize the data and to relate it to the characteristics of future projects. It is also necessary to time-stamp the data so that it will be possible to discard it when changes in technology or some other fundamental shift in the process makes it obsolete. Another important consideration when using historical data is to monitor the predictions trend to prevent dysfunctional projects from leading the organization productivity downwards, as shown by Figure 7.13.

7.2.3   Process improvement

Every process has a bottleneck that its process capability to its present level. The bottleneck could be an activity, a resource, or a policy. Its main characteristic is that any improvements that do not remove the constraints imposed on the process by the bottleneck will not have any effect on the overall R&D chain. Its second characteristic is that bottlenecks are move-able. As soon as the restrictions imposed by the current bottleneck are lifted, something else becomes a bottleneck [5].

Following this line of reasoning, measurement for process improvement should concentrate on two things:

1.   Identifying bottlenecks;
2.   Conveying an understanding of the forces and structures that shape the process so that it is possible to foresee the consequences to the overall process of removing the constraint.

Most capacity bottlenecks originate in mismatched resources' through-put and in the random variability intrinsic to the nature of development work. Examples of mismatched resources are too many researchers for the testing resources available, which forces researchers to line up for their turn in using the equipment, and too many requests for information for the number of customer support representatives. Examples of variability can be found everywhere. How long does it take to write a project specification?

Based on this, measurements for bottleneck identification should focus on the inter-arrival time between units of work, processing times, and their respective variability. The processing-time information could be derived directly from project data and the inter-arrival data could be extracted from a version control or document management system. When it comes to

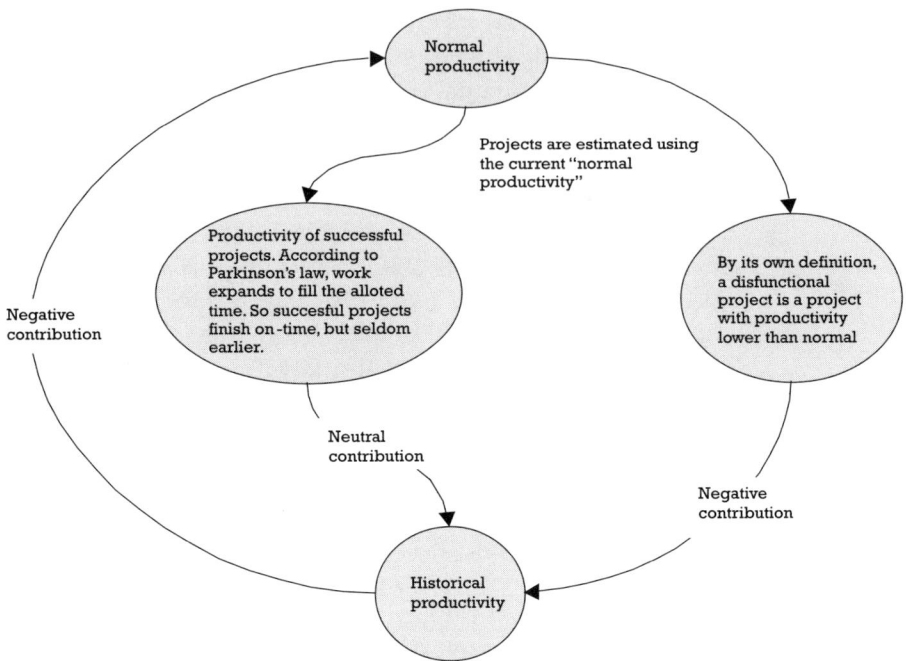

Productivity trend using a moving average of the 6 most recent projects

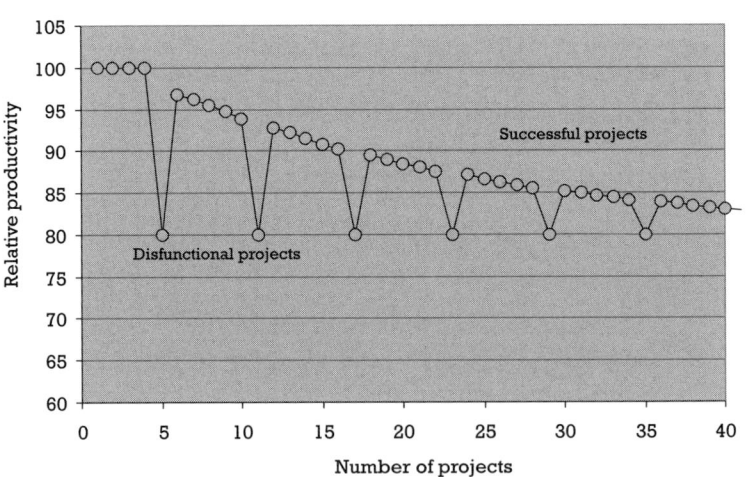

Figure 7.13 Danger of using historical data for estimation purposes. Normal productivity deteriorates over time.

quality, an approach like orthogonal defect classification (ODC) [6] is required to identify the process constraints. ODC's goal is to provide a multidimensional measurement scale for extracting key information from defects and enable the establishment of cause-effect relationships. Essentially ODC categorizes a defect into classes (see Table 7.1) that collectively point to the part of the process which needs attention, much like characterizing a point in a Cartesian system of orthogonal axes by its (x, y, z) coordinates. The basic ideas of ODC can be extended to other aspects of the process, such as change requests and the tracking of hours spent.

Once the process constraints have been identified, we need to give meaning to them; that is, to understand how they relate to other process parameters. One way to do this is to develop influence diagrams like the one used in Chapter 2 to describe the relationships between individual project delays and common management actions. An influence diagram (see Figure 7.14) is a tool for reasoning about systems. Whether a social or a physical system, the diagram shows the relationships between the variables that characterize and influence the system's behavior.

In most systems, complex behavior stems from two simple feedback loops, positive or self-reinforcing loops and negative or self-correcting loops. Positive loops amplify whatever is going on in the system and negative loops oppose or attenuate change (see Figure 7.15). Influence diagrams and their more quantitatively oriented counterparts, system dynamics diagrams, have

Table 7.1 ODC Classification Dimensions

	Time	Status	Impact	Phase	Trigger	Source
Definition	Time at which the event occurred	What is being done about it	Where the error would have surfaced during operational use	When the defect was introduced	What made the defect surface	Where the defect was found
Use	Measure Inter-arrival distribution and waiting time distribution	Measure queue length and assess project progress	Establish the benefits to be realized by preventing this type of defect	Identify activities that should be improved	Measure activities' effectiveness in finding this type of defect	What are the specific things to which we should be paying closer attention
Typical values	Dates at which the defect was reported, assigned, work started, fixed, closed	Reported Assigned In-progress Fixed Closed	Installation Security Performance Maintenance Serviceability Documentation Usability Customer expectations	Requirements Design Testing Integration	Review Inspection Test Field use	Documentation Own product Reused product Off-the-shelf third-party product Contracted third-party product

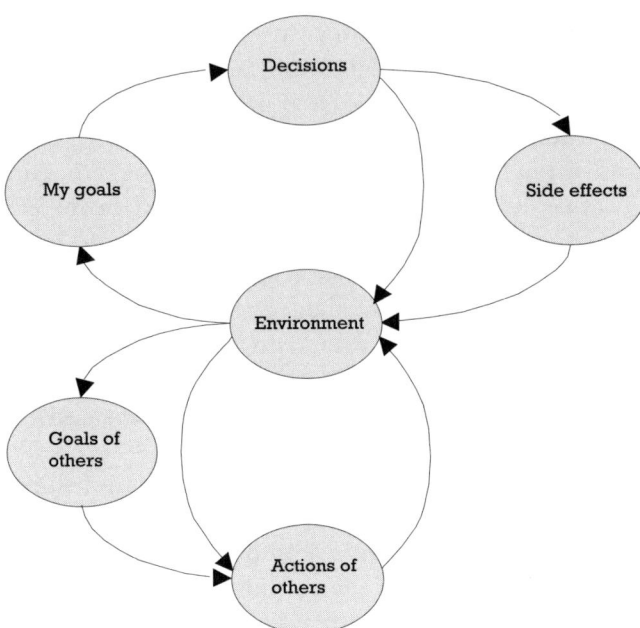

Our decisions alter the environment, leading to new decisions, but also triggering side effects, delayed reactions, changes in goals and interventions by others. These feedback loops might lead to unanticipated results and ineffictive policies.

Figure 7.14 Influence diagram showing effect of decisions.

been used to study project behavior and to recommend actions for at least 20 years. In his influential article, K. Cooper [7], used system dynamics to model the rework cycle in a project and T. Abdel-Hamid [8] (see Figure 7.16) proposed a complete model of the dynamics of a software project.

7.3 Selecting metrics

The purpose of measurement is to help management anticipate potential obstacles to success and, in the event that these potential problems turn into actual problems, to help decide on the best course of action. As the measures of success and the problems that the PO and the individual projects tackle are different, the metrics for the project-based organization must reflect this duality.

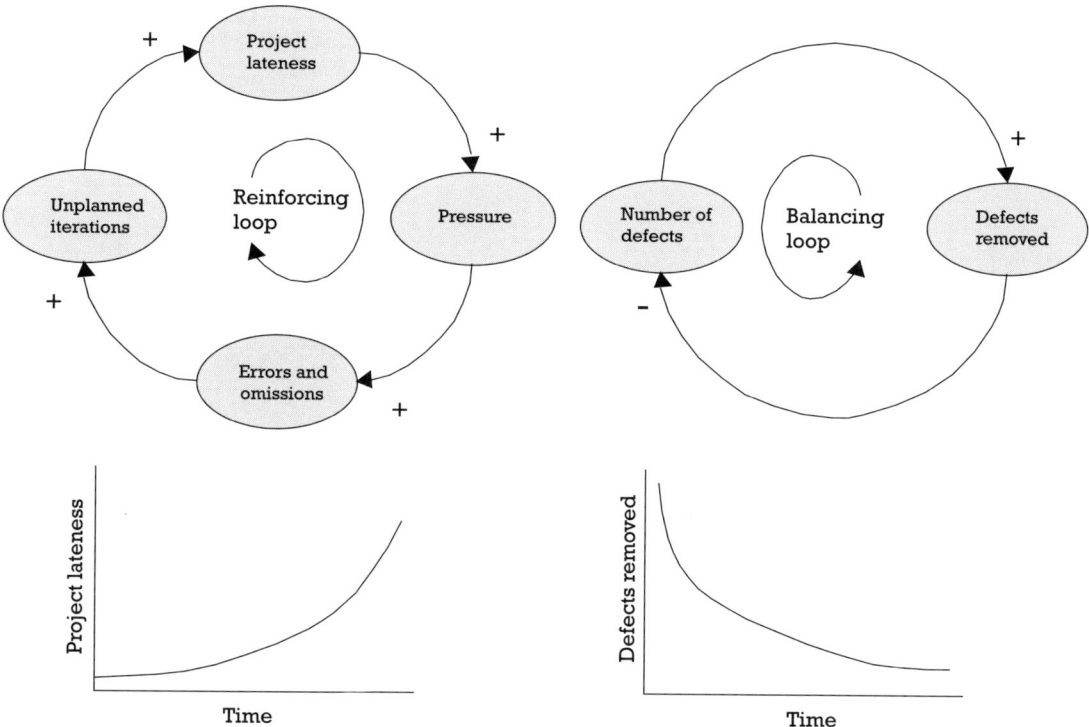

Figure 7.15   Positive and negative feedback loops.

While for the individual projects, the objective is to complete their work while meeting cost, schedule, quality, functionality, and technical perform- ance requirements, for the PO the objective is to maximize the benefit for the organization across all the projects and in the long run. To this effect, the PO needs to collect data not only to manage the project portfolio, but for estimation and process improvement purposes, none of which is an imme- diate concern for the individual projects.

The following sections describe some of the metrics commonly used in the project and the portfolio management environment. The list is necessar- ily incomplete and the reader is directed to additional resources; excellent material is available in [9] and [10].

7.3.1   Progress metrics

These metrics address the accomplishments of the project toward its final objective in relation to its planned time line.

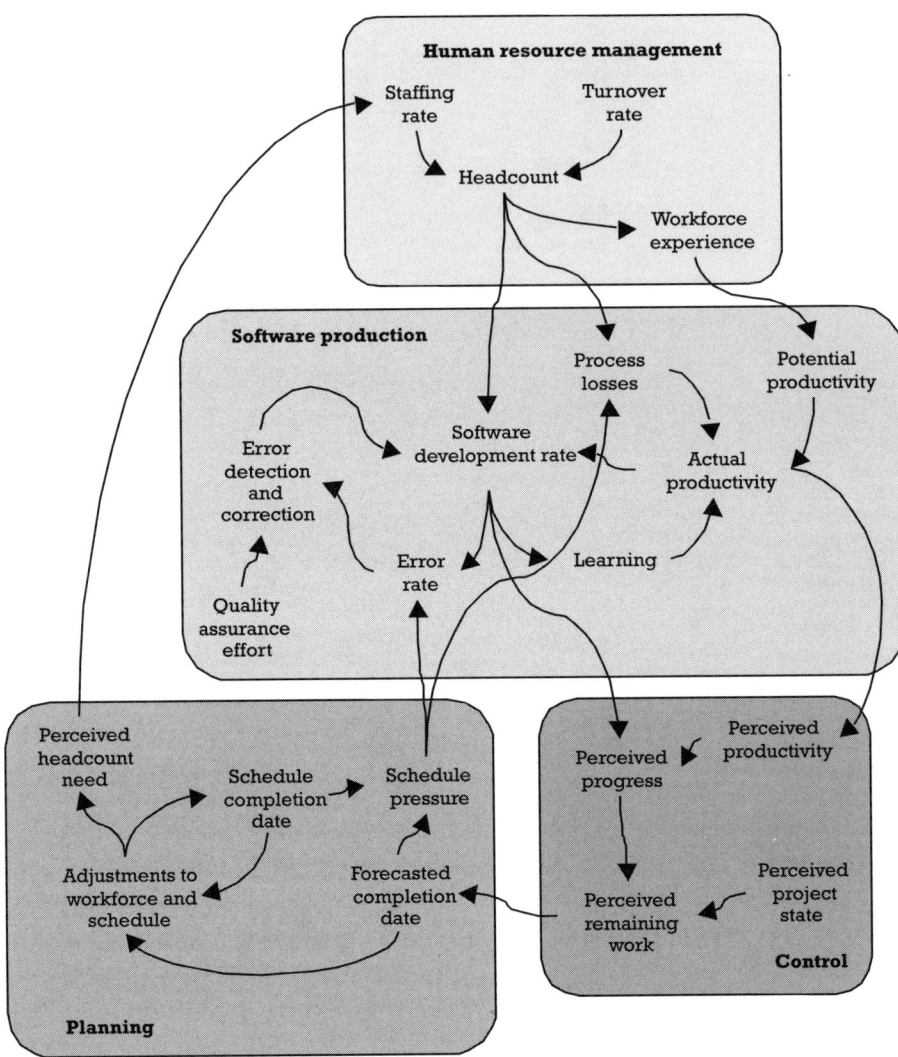

Figure 7.16 Model of software project dynamics showing relationship between different management variables.

The assumption behind all the progress metrics is that the future will look more or less like the past. In other words, progress is gradual, and although breakthroughs do occur, if you have not reached what you were set to achieve by a certain time it is unlikely that you will beat all the odds and be able to finish your work in the planned time. Table 7.2 shows some typical metrics for this area.

Table 7.2 Typical Progress Metrics

Category	Metrics	Description
Work progress		The progress of a specific task could be measured in terms of its main output (i.e., requirements defined, LOC, errors found, pages of documentation written, etc.)
	Requirements activity	Number of new requirements specified during the reporting period.
	Source code activity	Number of lines of code added, changed, or deleted during the reporting period.
	New trouble reports (TRs)	Number of newly written TRs during the reporting period
	Closed TRs	Number of TRs solved during the reporting period
	Open TRs	Number of unsolved TRs during the reporting period
Technical performance measures		TPM are deliverable specific metrics that track design progress toward meeting customer performance requirements. The technical parameter to be measured depends on the type of deliverable being developed, so here we limit ourselves to cite a few examples. A TPM should be a significant qualifier of the total system, and reflect a characteristic that contributes to system success. Critical technical parameters can be derived from identified risks, system requirements, safety issues, cost/schedule drivers, and mission parameters.
	Weight	TPM attribute. Example: an aircraft, where vehicle weight is critical to range and flight economy.
	Transactions per second	TPM attribute. Example: an automated reservation system, which should be capable of dealing with thousands of information requests per second.
	Mean time between failures (MTBF)	TPM attribute. Example: a telecommunications switch in a telephone network.

7.3.2 Performance metrics

These measurements address the amount of resources (i.e., time and money) spent to attain the present level of achievement in relation to the planned expenditures or to target ratios in the organization.

Probably the best-known exponent of this type of measurements is earned value. Table 7.3 shows another common metrics

Figure 7.17 shows an analysis of the significance of several cost performance and schedule performance curves. Performance measurements do not tell what is happening with the project, only that things are not going according to plan (whether for good or bad) and should be looked at. Performance measurements are not capable of discriminating between a potential problem originated in the project's execution or a latent one hidden in its plan.

Table 7.3 Typical Performance Metrics

Category	Metric	Description
Earned value		At the most basic level metrics in this category compare the actual cost of the work performed to the budgeted or planned cost of that same work and to the budgeted cost of the work that was scheduled to derive the four measures described below.
	Cost variance	Measures the difference (positive or negative) between the actual and the budgeted cost of the work performed.
	Schedule variance	Measures the difference (positive or negative) between the budgeted cost of the work performed and the budgeted cost of the work scheduled.
	Cost performance index	Is the ratio of the budgeted cost of the work performed to the actual cost of the same work. An index value of 1 indicates that the project spending is proceeding according to the plan. An index below 1 indicates overspending. An index over 1 indicates that work is progressing at a lesser cost than planned.
	Schedule performance index	This is the ratio of the budgeted cost of the work performed to the budget cost of the work scheduled. An index value of 1 indicates that the project is progressing according to the plan. An index below 1 indicates delays. An index over 1 indicates that work is progressing at a faster pace than planned.
Productivity		Productivity is the ratio of the amount of product or output of the organization relative to the resources consumed to produce it. There are several options for measuring output. It can be measured in terms of number of products fielded, number of features delivered in the products, or some measure of the size of the products such as lines of code (LOC) or function points (FP). Resources consumed will most likely be represented by effort expended, measured in terms of hours or hours per month.
	SLOC-man month	Measures the average amount of software produced, in source lines of code (SLOC), by person by month. Usually is compared to past performance.
	Multifactor productivity (data envelopment analysis)	In the case of multifactor productivity, instead of a single input like man-months, the resource consumed is a composite of labor, capital investment, and other resources used in the process.
Ratios		A ratio is simply a number expressed in terms of another, and is used to restate the relative magnitude of the dividend to the divisor using a single number.
	Rework ratio	The rework ratio measures the amount of work effort expended to fix defects in relation to the total work. Rework may be expended to fix any product. This measure identifies the quality of the initial project effort, products that need the most rework, and processes that need improvement.
	Financial ratios	Ratios such as profit margin and return on investment (ROI) relate the benefits generated by a project to the amounts invested to obtain them.

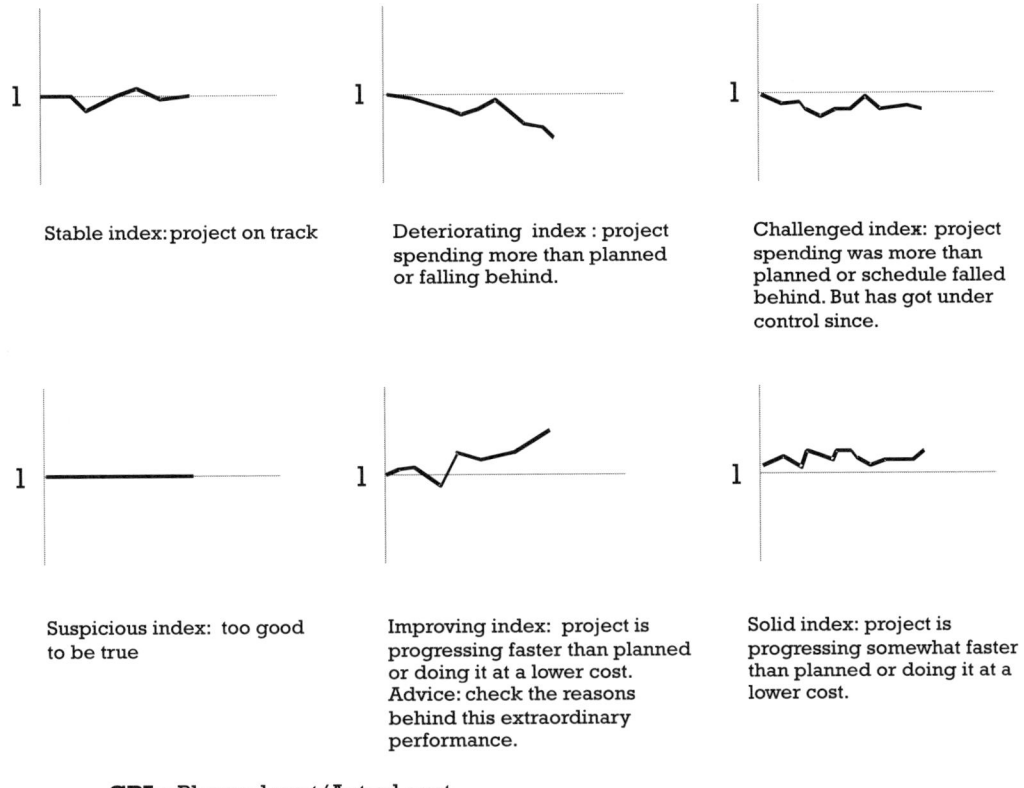

Stable index: project on track

Deteriorating index : project spending more than planned or falling behind.

Challenged index: project spending was more than planned or schedule falled behind. But has got under control since.

Suspicious index: too good to be true

Improving index: project is progressing faster than planned or doing it at a lower cost. Advice: check the reasons behind this extraordinary performance.

Solid index: project is progressing somewhat faster than planned or doing it at a lower cost.

CPI = Planned cost/Actual cost

CSI = Planned cost of work accomplished/Planned cost

Figure 7.17   Interpreting cost and schedule performance indexes.

7.3.3   People

These measurements relate to the level and adequacy of the staff allocated to the project and to the amount of overtime, direct hours, and employee turnover attributable to a project. See Table 7.4 for some examples of people metrics.

While the reasons for tracking the actual staffing levels against the planned ones are obvious, the motivation behind some of the other metrics needs to be analyzed. In Chapter 2 we saw the negative consequences of overtime and fatigue on quality and productivity; furthermore, if key people leave in the middle of the project, or if half the staff is alienated, the damage could be insurmountable for the project and far-reaching for the organization (see Figure 7.18) for an example of staffing charts.

Table 7.4 Typical People Metrics

Category	Metric	Description
Staffing		Measures in this category are used to evaluate the adequacy in terms of number and experience of personnel assigned to a project and the level of stress or morale of the staff.
	Staffing variance	Shows the difference between the staff required by the current plan and the allocated head count.
	Slack index	This index measures the degree of freedom available for innovation, for knowledge sharing, and for personal and organizational development. Indirectly it measures the margin of maneuver the organization has to respond to a surge in work or an emergency condition.
		Although the need for some "slack" is acknowledged in planning constants, which allocate less than the totality of the available work hours to direct tasks, this measure is seldom tracked.
	Overtime index	Overtime is not only expensive, but it is unproductive and harmful if abused.
		Overtime is a leading indicator of unsound working conditions. Calculated by dividing the overtime hours by the base working hours for all project staff in this reporting period, it is expressed as a percentage. The target range is less than 10%. When the overtime rate approaches 20%, the ability of the staff to respond effectively to crises suffers significantly.
	Voluntary turnover	Each project member who leaves the team causes a productivity drop and schedule disruption. A high turnover rate could be indicative of a morale problem, excessive pressure, etc.

Long overtime hours and the lack of any "slack" in the employee work-week are leading indicators of trouble ahead. If a project can only meet its commitment by resorting to continuous overtime and the postponement or abandoning of training and other activities, the sustainability of such a pace shall be brought into question not only from an employee morale point of view but also from a cost perspective.

The employee turnover rate for the project should also be compared with that of the entire organization and other projects for signs of trouble. A project that a large number of people are trying to leave is not a healthy project.

To prevent overreacting to an increase in overtime or employee turnover, it is important to distinguish between random variations in weekly reports and institutional trends. The best way to do this is to use control charts similar to those used in statistical process control (SPC) (see Figure 7.19).

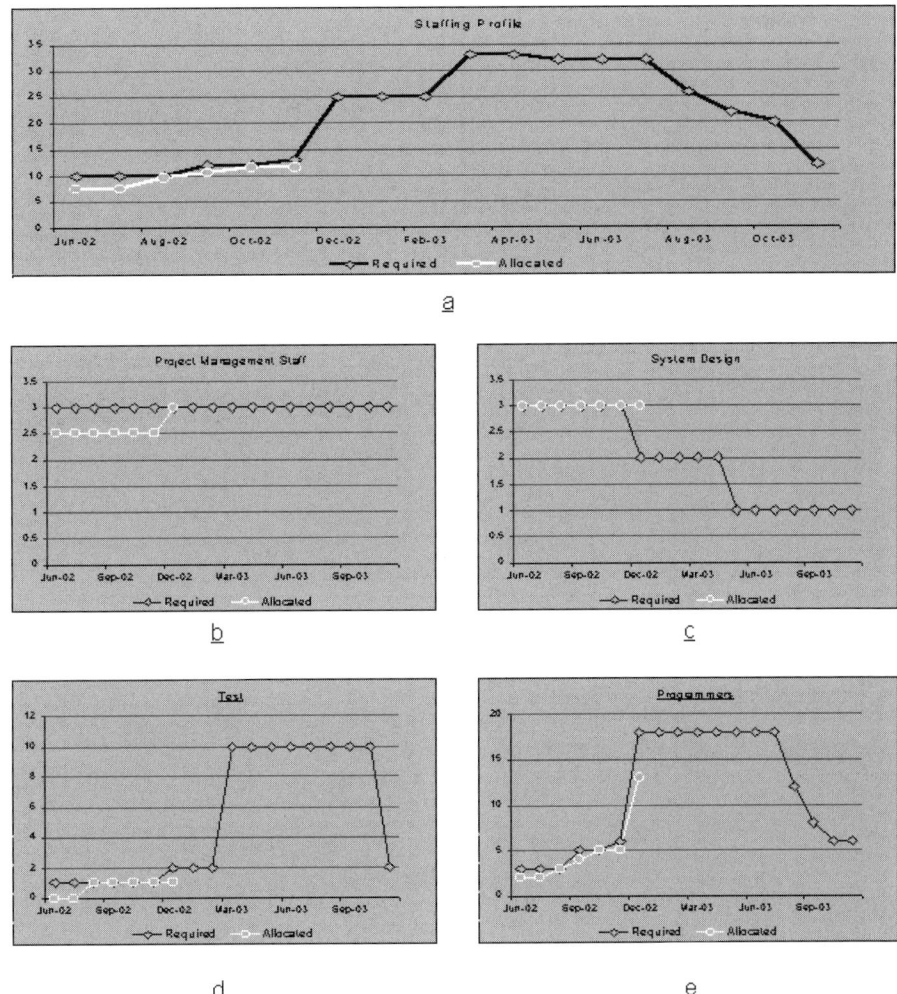

Figure 7.18 Staffing profiles: (a) aggregated across all competence areas; (b–e) for each competence area. This level of detail is necessary because the aggregated curve could mask surplus in one area with shortfalls in another, which are not interchangeable.

7.3.4 Product

Few things could cause more damage to a project than the lack of stable requirements or changing interfaces. This set of measurements (see Table 7.5) addresses the amount of change or increase in the project's scope of work and technical interfaces.

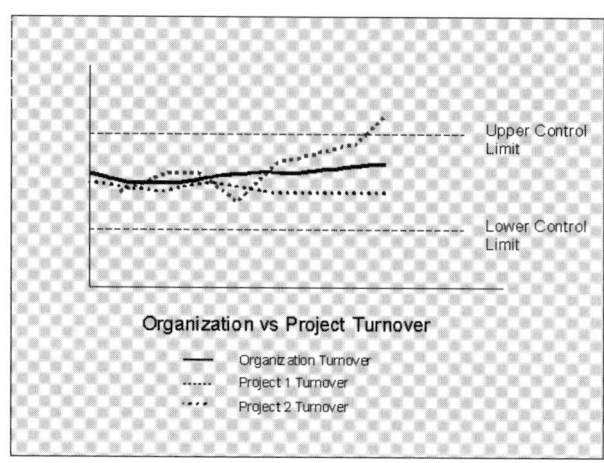

Figure 7.19 Control chart assessing project morale. Project 1 seems to be experiencing a larger-than-expected turnover. This might be indicative of morale problems within the project.

Table 7.5 Typical Product Metrics

Category	Metric	Description
Requirements	Requirements growth	An inordinate number of new requirements might be a sign of an increasing work scope that was never planned for.
	Requirements churn	A high rate of requirements change can indicate that the customer is not sure of what is wanted, or that the original requirements definition was poor. A high rate often predicts disaster for most projects.
Interfaces	Interfaces churn	An inordinate rate of change of the project interfaces signals an unstable project environment or a very bad design.

7.3.5 Product quality

Quality in this context refers to the ability of the project deliverable to meet its stated requirements. The reason for taking such a narrow view of quality is that the measurements in this section are used mainly to determine whether or not the project deliverable is ready to be released to the market, transferred to manufacturing, and so on (see Table 7.6). The broader issue of customer satisfaction is addressed in the next section.

Table 7.6 Typical Product Quality Metrics

Category	Metric	Description
Functional completeness		The percentage of functional, maybe weighted, requirements implemented versus those defined for the deliverable.
Performance		The percentage of performance, maybe weighted, requirements implemented versus those defined for the deliverable.
Usability		The percentage of usability, maybe weighted, requirements implemented versus those defined for the deliverable.
Reliability		These types of measures are designed to give an indication of the level or continuity of service that could be expected from the delivered product.
	Open problems	This measure quantifies the number, status, and priority of the problems reported. The quantity of problems reported is correlated to the amount of rework necessary before the system can be released for full operational use. Arrival rates could be used to determine the worthiness of continuing testing, while closure rates could be used as a predictor of testing completion.
	Fault density	Expresses the ratio of the number of problems written against a component relative to its size. It is useful to compare the relative quality of different components.
	MTBF	Indicates the amount of time a product can be expected to function without experimenting a failure.
	Availability	Indicates the percentage of time that the product is available for use. Is the result of two main parameters: MTBF and mean time to repair (MTTR)

7.3.6 Customer satisfaction

Customer satisfaction is defined as how clients perceive and judge the project's attempt to satisfy their needs and fulfill their expectations. Customer satisfaction could be measured directly through questionnaires or indirectly using measurements such as number of items returned and repeated sales. These measures are important to the long-term sustainability of the organization. See Table 7.7.

7.4 Summary

One of the most important elements of problem solving is information. Without the proper information, problems cannot be solved to any degree of accuracy. Solely using gut feelings to solve problems can often cause more problems than are solved. We need to measure in order to do the following:

Table 7.7   Typical Customer Satisfaction Metrics

Category	Metric	Description
Deliverable		The project's deliverables should be evaluated at least from the following six perspectives.
	Functionality	This aspect of customer satisfaction refers to the ability of the product/deliverable to fulfill the customer needs.
	Reliability	This refers to the absence or presence of failures that prevent the customer from enjoying the functionality to be provided by the deliverable.
	Maintainability	How easy is to fix or upgrade the deliverable?
	Performance	How long? How many? How often? At which rate?
	Cost	Costs what the customer and the seller agreed it should cost as expressed in a project plan or in updates to it?
	Delivery expectations	Is delivered in accordance with schedules agreed to by the customer and the seller in a project plan or in updates to it?
Support		This activity takes place after the output of the project has been delivered, and in consequence, it is not measured within the project itself.

- Know where we are in relation to where we were supposed to be;
- Know trends;
- Predict future status;
- Facilitate comparison and benchmarking;
- Plan new projects;
- Understand and model the impact of driving factors behind performance;
- Find and give priority to improvement actions;
- Verify effects of actions and relate these effects to goals.

Measurements and models by themselves do not result in good decisions. They do not replace thinking, knowledge, and good judgment, but they do provide the objective foundation that good project managers need in order to make quality decisions.

References

[1]   Weinberg, G., *Quality Software Management: Systems Thinking*, Vol. 1, New York: Dorset House Publishing, 1992.

[2]   Trochim, W., *The Research Methods Knowledge Base*, Cincinnati: Atomic Dog Publishing, 1999.

[3] El Emam, K., et al., "The Confounding Effect of Class Size on the Validity of Object-Oriented Metrics," *IEEE Trans. on Software Engineering,* Vol. 27, No. 7, July 2001.

[4] Drucker, P., *Management Tasks, Responsibilities, Practices,* New York: Harper and Row, 1973.

[5] Goldratt, E., *The Haystack Syndrome: Sifting Information out of the Data Ocean,* Great Barrington, MA.: North River Press, 1991.

[6] Chillarege, R., "Orthogonal Defect Classification: A Concept for In-Process Measurements," *IEEE Trans. on Software Engineering,* Vol. 18, No. 11, November 1992, pp. 943–56.

[7] Cooper, K., *The Rework Cycle: How Projects Really Work, and Rework,* PM Network, PMI, February 1993.

[8] Abdel-Hamid, T., *Software Project Dynamics: An Integrated Approach,* Upper Saddle River, NJ: Prentice Hall, 1991.

[9] McGarry, J., et al., *Practical Software Measurement: Objective Information for Decision Makers,* Foundation for Objective Project Management, Boston, MA.: Addison-Wesley, 2001.

[10] Ellis, L., *Evaluation of R&D Processes: Effectiveness Through Measurements,* Norwood, MA: Artech House, 1997.

CHAPTER

8

Contents

Deploying the project office

Since many of the problems experienced by the organization today are to a large extent the result of actions it sanctioned and behaviors it rewarded in the past, PO deployment requires that, in addition to processes and tools, the organization revisit the assumptions on which it was built and is now being run.

Very few of the management "silver bullets" that were proposed in the last decade are still around, in general not because of their lack of merit, but because they ignored the culture of the organizations in which they had been deployed. A lasting and effective change therefore requires the alignment of the organization's culture, its reward system, power structure, and mental models in consonance with the PO's primary rationale for its existence: maximizing the benefits across the entire portfolio, rather than on individual projects. It requires that the organization stop rewarding employees who "fix" problems and start rewarding those who do not create them. It requires that the organization stop promoting staff members who promise things that they later cannot deliver and start promoting those who take calculated risks.

8.1 Layers of change

Most change and improvement efforts live short lives or fall short of the expectations they were born with. The reason for this is that most organizations make only the relatively easy changes: They introduce a new process, a new tool, maybe even a reorganization of sorts, but they fail to make the deeper cultural, political, and behavioral changes needed to institutionalize the new mindset and ensure the realization of its long-term benefits [1].

Effective and lasting changes require that the organization make congruent decisions in each of the nine areas of concern depicted in Figure 8.1. By congruent decisions, we mean decisions that support or reinforce one another and not decisions that by design or omission promote conflicting behaviors, because as Figure 8.2 shows, contradictory messages do not result in a balance between competing objectives but in a loss of overall performance and dysfunctional behaviors across the organization.

8.1.1   Reward structure

Promotions, incentives, and recognition are the mechanisms that organizations use to further their organizational goals by matching individuals to the jobs they are best suited for and enhancing the motivation and satisfaction of individual employees. Promotions serve two important and distinct purposes. First, individuals differ in their skills and abilities and jobs differ in the demands that they place on individuals. Promotions are a way to match individuals to the jobs for which they are best suited. The matching process occurs over time as employees accumulate intellectual capital and as more information is generated and collected about their talents and capabilities. A second role of promotions is to provide incentives for employees through the pay and prestige associated with a higher rank in the organization.

While promotion relates to the long-term relationship between the employee and the organization, incentives and recognition focus on promoting specific types of behaviors within definite time frames. Incentives

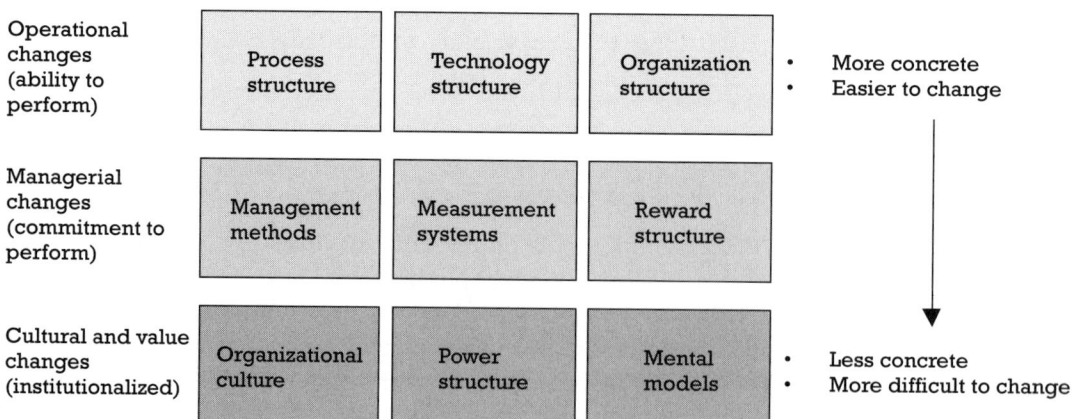

Figure 8.1   Business-process-improvement layers of change. (*After:* [2].)

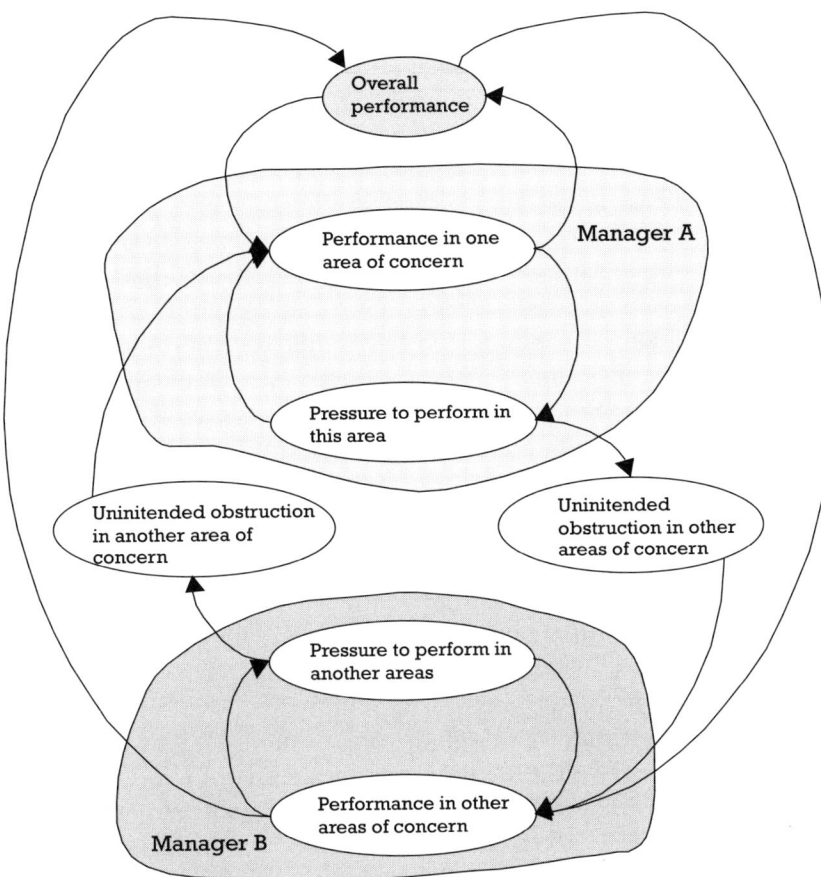

Figure 8.2 Incongruent messages and their negative effect on overall performance. (*After:* [3].)

such as a pay raise, bonus, and stock options have a monetary value, whereas recognition in the form of awards for good performance and favorable mentions in company newsletters are symbolic rewards, satisfying socioemotional needs. Some other rewards, such as dinners for two or tickets to sporting events, although they have a monetary value, are also primarily symbolic.

Pay-for-performance systems were touted as the answer to linking contributions to rewards, but since their implementation is usually plagued with unanticipated side effects, their efficacy in promoting sought behaviors is far from being universally accepted. There is, however, total agreement that a misaligned reward system guarantees dysfunctional behavior.

To align the long-term view that should prevail among the PO staff with the reward system, small annual rewards should be replaced by more significant ones, to be paid once the results, desired or otherwise, of a given administration have had the time to manifest.

Rewards should also be the result of a multidimensional evaluation, because single objectives, achieved at the expense of other equally important concerns, are easy to attain. For example, what good does it do an organization to deliver a project on time if half the members of the project, including the organization's best designer, resigned in the process? What good does it do if a project is delivered on time, but the organization loses a client because of the product's lack of quality? What good does it do an organization to sell hundreds of systems if it is losing money on each of them, and worse yet, doesn't know that it is losing money?

8.1.2 Organizational culture

Organizational culture is the pattern of norms, values, beliefs, and attitudes that influence individual and group behavior within an organization. Originating with the founders of the organization, culture is shaped and honed over time by succeeding senior executives and other stakeholders. Culture filters down through the organization and is further refined and modified in the day-to-day priorities and actions of everyone in the business. Culture affects performance because it affects how people think, feel, and act and helps to determine the situations in which they act.

Enculturation into the beliefs, practices, values, and style of discourse of the organization occurs through work routines and positive reinforcement from someone who has already been successfully enculturated. Enculturation is not just the process of internalizing the knowledge and skills required by a job; it is the process of becoming a member of a community. For newcomers to the organization, enculturation entails picking up the relevant social language, imitating the behavior of successful members, and gradually beginning to act in accordance with community norms.

Organizational cultures fill the information gap existing between what is explicitly communicated and what is required by the task at hand by providing preexisting ways of understanding what is occurring, how to evaluate it, and what kind of actions constitute an appropriate response to the situation. The desirability of a strong culture depends on the accuracy and validity of the knowledge it provides to the employees.

Among the mechanisms by which the PO manager could steer the evolution of a project management culture within his or her sphere of influence are as follows:

- A clearly articulated vision, expressed in the PO mission and its processes;

- The recruitment of like-minded employees;

- The use of symbols to reinforce cultural attributes;

- Repetitive socializing and training of employees;

- The praise and reward of behavior consistent with the desired culture;

- The design of an organizational structure that reinforces the desired values.

Having a strong culture could be a detriment, if as in the case of the now defunct People Express [4] it prevents the organization from learning and adapting to a changing environment. People Express Airlines was patterned after the values of Don Burr, its founder and chief executive officer, who effectively used cultural levers to develop a strong company culture. Burr's explicit purpose was to form an airline that would be a model of customer concern, people sensitivity, and teamwork. People Express achieved almost unbelievably successful results during its first 5 years of existence, setting world records for income and profitability. However, a change in environmental demands brought about by the airline's purchase of Frontier Airlines, a unionized company, led to the rather swift demise of both companies. People Express, with its strong culture, was simply unable to adjust to the requirements of a radically different environment.

8.1.3 Power structure

According to Rosabeth Kanter [5], power is the United States' last dirty word. It is easier to talk about money and much easier to talk about sex than it is to talk about power. People who have it, deny it; people who do not have it, do not want to appear hungry for it; and people who engage in its machinations, do so secretly.

Robert Dahl [6] has defined power by saying that person A has power over person B to the extent that A can get B to do something that he would not otherwise do. Access to resources, information, and the ability to show discretion or exercise judgment in extraordinary circumstances are attributes of organizational power. Because of this, people with power are capable of accomplishing more and passing on more resources and information to subordinates. When employees regard their manager as powerful or influential, they see their status enhanced by association and they generally have high morale and feel less critical of or resistant to their boss. More powerful leaders are also more likely to delegate—since they are too busy to do everything by themselves—to reward talent, and to build a team that places subordinates in significant positions. The result of power is then more power.

Although people vary to the extent that they seek power, they rarely relinquish it voluntarily. This is eloquently expressed by Machiavelli in *The Prince* [7]:

> It ought to be remembered that there is nothing more difficult to take in hand, more perilous to conduct, or more uncertain in its success, than to take the lead in the introduction of a new order of things. Because the innovator has for enemies all those who have done well under the old conditions, and lukewarm defenders in those who may do well under the new.

Power in organizations comes from a variety of sources. Structural sources of power refer to the power that the organization vests in an individual based on the work or position he or she is responsible for. Personal power is derived from the characteristics of the person him or herself, such as expertise or friendship. Structural sources of power include the following:

- Reward power (the capacity to dispense rewards);
- Coercion power (the capacity to dispense punishments);
- Task power (the power that accrues naturally to a particular role in the organization);
- Legitimacy power (the authority that emanates from a position in the organization).

Personal sources of power include the following:

- Expertise power (the power derived from the possession of valuable information or status);
- Referent power (the power arising from the desire of others to imitate or be agreeable to the referent).

The introduction of a PO into an organization implies a shift in the balance of power. The design and location of the PO manager job makes it a very desirable position for anyone aspiring to an executive role in the organization. Not only is the job relevant, it allows considerable discretion in its exercise and spurs close contact with higher-level people who confer approval, prestige, and recognition. PO power comes at the expense of the line functions, which formerly oversaw the execution of the projects, the sponsors, accustomed to dealing directly with project managers with little or no oversight, and the project managers, who might see their margin of maneuver reduced. The paradox is that these are precisely the people the PO manager needs to support the change. Despite what much of the literature says, operational changes do not take place at the top nor at the bottom. They take place at the middle. If the PO concept is going to work, it is the line managers, the product managers, and the project managers who need to buy into it. Therefore, it is crucial that the people responsible for the deployment of the PO acknowledge the existence of these concerns and make sure that the changes introduced do not render other levels of the organization powerless.

Power is distributed across the organization and not just concentrated at the top, so even apparently powerless members of the organization who feel threatened by change can retaliate. This is clearly illustrated by the relationship between a manager and a subordinate, where the manager who asks the subordinate to do a particular job becomes dependent upon the subordinate to complete the task correctly and on time. Certainly the manager has the power to replace the subordinate if the task is done poorly or late, but firing the subordinate will not change the fact that the task needs to be done and that valuable time has been lost. Furthermore, the manager will not escape the situation unharmed; to some extent he will be hurt by his inability to get the job done without resorting to the power of his position.

People do not resist change because they are stubborn or because they are afraid of new things, they resist change for reasons that make good sense to them even if they do not make sense to us. Resistance is seldom overt, it rather manifests in the form of pseudo-technical excuses explaining why the PO, although a good idea, will not work in this particular organization. It manifests also by noncooperation and misinformation that could lead the PO manager to make some bad choices. In making changes, then, it is wise to make sure that the people affected are involved, informed, and taken into account, so that the process can be used to build their own sense of worth. If such involvement is impossible, then and only then is it time to ask for managerial support to move these people out of the territory altogether.

8.1.4 Mental models

Aside from organizational culture and political structure, employees have their own mental models through which they perceive and interpret the messages generated by the organization.

There is a simple experiment, popular among organizational behavior professors to illustrate the concept of mental models, in which a group of students is asked to read a short and simple story, around two pages long, and then to rate the characters appearing in it from most despicable to least despicable and then explain these ratings. The surprising thing about this experiment is that rarely do two students come up with the same ranking for the characters, and that some of the traits that cause people to pick one character over another are completely ignored by other participants reading exactly the same material.

Mental models can range from simple generalizations, such as "people work best under pressure," to complex models such as the one presented in Chapter 2 to explain the propagation of delays across projects. But what is most important is that mental models are active, they shape what we see, how we interpret what we see, and how we respond to it (see Figure 8.3). The perceiver selectively attends to sensory inputs, constructs a representation of the inputs, and attaches a meaning to the constructions. Perceptions

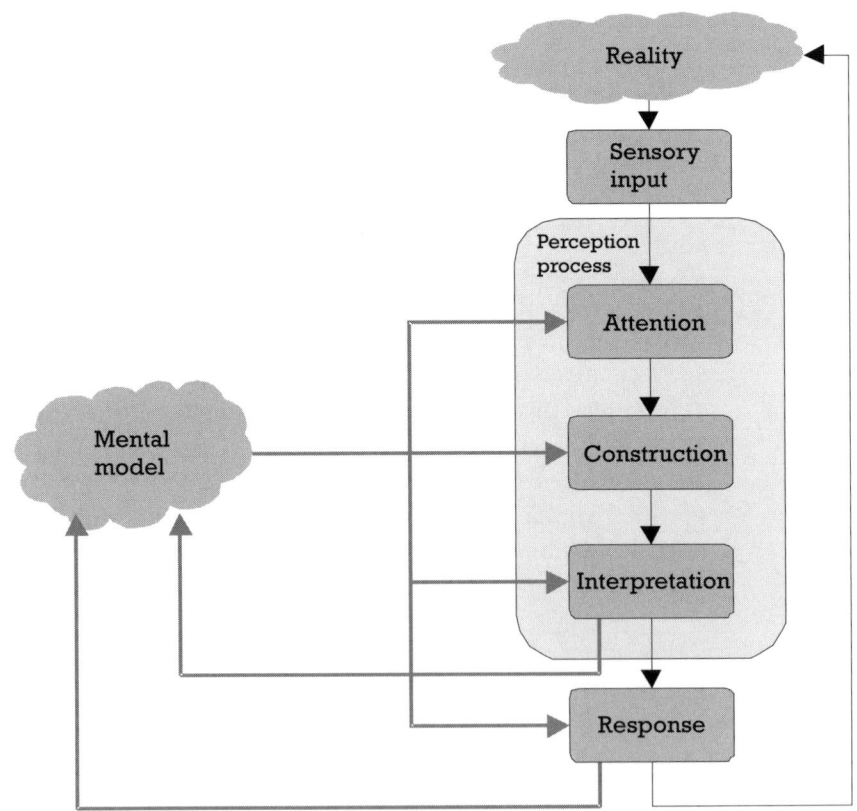

Figure 8.3 How the mental model conditions the active process of perception. (*After:* [8].)

are then used as inputs to elaborate responses that again are conditioned by our beliefs about how the world works.

Mental models also play a role in filling in gaps of missing information required to make a decision, so in the absence of knowledge or facts, we use beliefs. If our mental model holds that people work best under pressure, when we are confronted with a project that is slipping away, we are going to conclude that we are not exerting enough pressure. Furthermore, we are likely to look, and find, clues that confirm that this is in "fact" the case and disregard or explain away contrary evidence.

In his book *The Fifth Discipline* [9], Sengue explains that the first thing that needs to be answered about mental models is not whether they are right or wrong, obviously a fairly important thing, but rather whether they are tacit or explicit. A tacit model is a model that exists below the level of awareness; in consequence, it prejudices our judgment without us knowing it.

As the stock market demonstrated during the dot-com era, entire industries can develop chronic misfits between mental models and reality. But as devastating as mental models can be to any change effort, they can also be powerful change accelerants if they are made explicit and discussed. Shell and the World Bank have used this approach for internal improvement processes as well for initiating change in their external projects.

8.2   Where to start

The assessment of the current situation and an educational campaign are always good starting points. Even if you think you know what the problems are and how to address them, you need to begin with an assessment of the current situation. Besides the obvious gathering of information on the current strengths and opportunities for improvement, the undeclared purpose of the assessment is for all the stakeholders to achieve a common understanding of what the problems are and what can be done about them.

Achieving a common understanding of what the problems are is not a minor accomplishment. As mentioned with respect to mental models, one should never assume that everybody sees the same problems or confers them the same significance.

The charts in Figure 8.4 show the disparity of responses given by the senior executives and project managers of three global organizations to the following three questions:

▸ Do you have problems coordinating your project portfolio?

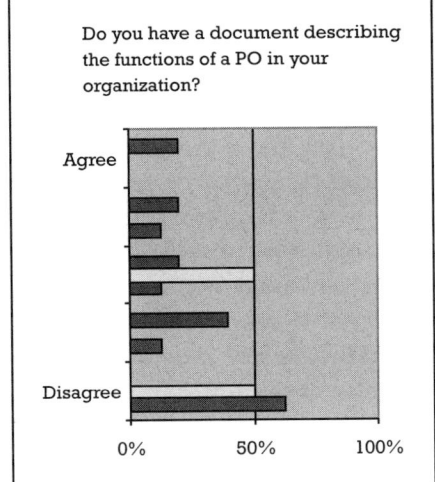

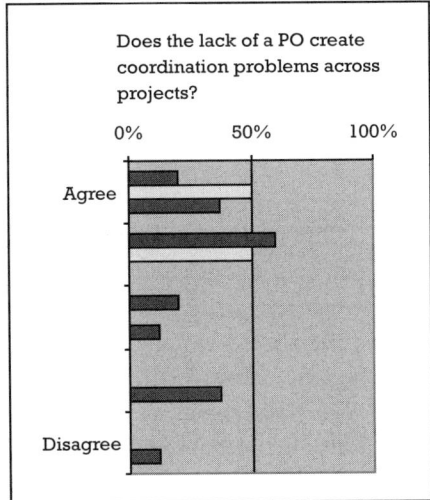

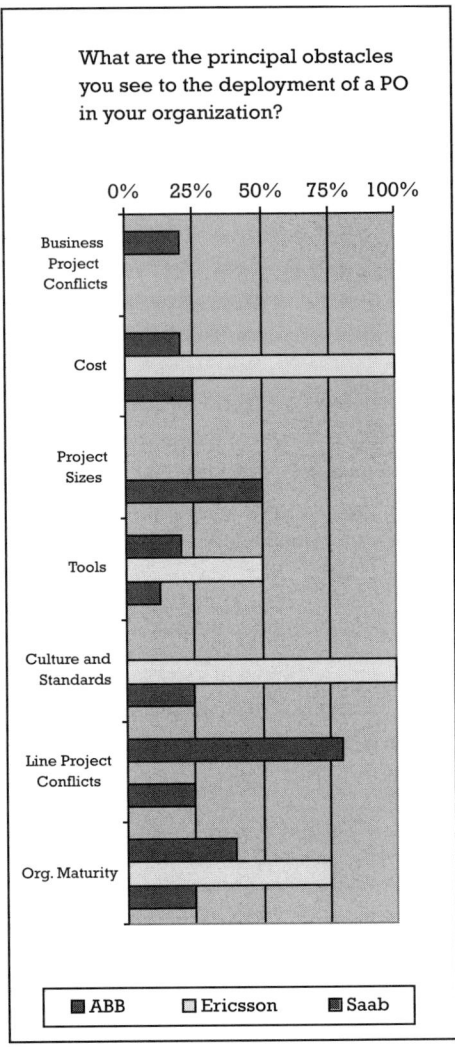

Figure 8.4 Different perceptions of realities at ABB, Ericsson, and Saab. (*Source:* [10].)

  ‣ Do you have a document describing the functions and role of a PO?

  ‣ What is the most important obstacle in deploying a PO in your organization?

The disparity of responses within a given organization reflects the different understanding and perceptions of the respondents. The lack of a common understanding about the problems the deployment of a PO should solve could not only undermine the efforts but also result in missed opportunities if the alternative diagnostics turn out to be true.

Don't assume that everybody is at the same level of understanding or has the same knowledge you have. Albert Einstein once said, "It's not that I'm so smart, it's just that I stay with problems longer." If you have spent some time thinking and reading about the problems of multiproject management, it is highly likely that you have insights into the problem that people who have not done so do not have. This could be addressed by means of an education campaign aimed first at showing the problems associated with multitasking, coordination across multiple projects, the use of overtime, and so on. Whenever possible and avoiding assessments that sound like criticism, the campaign should be linked to actual problems experienced by the organization. Once people are aware of the problems and the mental models have started to surface, the organization is ready to move to the next stage.

8.3 Incremental deployment

You don't want to spend 18 months discussing how to deploy a PO in your organization, because in that time you would have lost your opportunity and the naysayers would have demonstrated that nothing but their way of working actually works.

An incremental deployment methodology is a deployment approach in which the total change project is divided into a series of short, intensive cycles of implementation, each of which delivers a tangible business benefit. There are several reasons for recommending an incremental approach rather than an all-at-once deployment. First, project managers are doers. They enjoy action and they want to see results. You are not likely to keep any momentum, nor raise any excitement, by talking about processes for a long period, so you need to scope the amount of change to be introduced to fit their attention span. Second, any innovation or change requires an assimilation period. The incremental approach provides the time necessary for this process of organizational learning to take place between consecutive deployment increments. If this time is not allowed, as in the case of all-at-once deployments, it is likely that many changes will pile up, one over the next, leading to frustration and rejection of the initiative under the pretense conveyed by the line "we are too busy to change." Third, an incremental approach prevents the common tendency to overengineer technology solutions while substantially shortening the time to the arrival of business benefits.

The first thing to do is to produce a blueprint of what the PO would look like: What responsibility would it have, what type of culture will it foster, what kind of tool or tools would be used? The level of detail of the blueprint should be such that it allows the purpose and operation of the PO to be communicated to management and other stakeholders and that it provides enough contextual information so that consistent decisions can be made all along the way.

The blueprint is also an excellent technique for managing expectations. The blueprint will allow you to disarm those that might paralyze the project with untimely demands about one functionality or another. The blueprint will allow you to respond to the requests by saying either that the requests have been considered and will be done in due time, or that they will be incorporated into the blueprint for prioritization.

Once the blueprint has been drawn, it is time to decide which tool or tools will be used to support the PO processes. Many would argue that in order to decide which tools to use, it is necessary first to define the processes they will have to support to the last level of detail. My experience is that tools such as the ones described in Chapter 5 provide a generic platform for portfolio management that could be adapted to many different processes; furthermore, a tool that cannot be adapted should be discarded automatically, since a tool that fits today's processes like a glove might not be useful for tomorrow's processes. The reason to move swiftly on this is that most of the processes need automated support. If everything needs to be defined before proceeding to tool selection, there will be an impasse of 6 to 12 months before the deployment will start.

With the blueprint completed and the decision on the automation under way, it is time to prioritize the order in which things will be done. The order of implementation should take into consideration the following:

- Business priorities;
- Technical dependencies;
- Political needs.

The guidelines listed below [11] provide advice on how to approach the planning and management of the deployment project to prevent "improvement burnout" and paralysis-by-analysis:

- Use concrete business objectives to drive the prioritization of the implementation process.

- Divide the implementation into a series of nonoverlapping increments, each of which enables tangible business improvements even if no further increments are implemented.

- Ensure that each increment implements everything required to produce the desired results (i.e., software functionality, policies, processes, training, and measures).

- Scope the increments so that each can be implemented in no more than 4 months.

- Use the results of each increment as a basis to adjust the blueprint and plan for the next increment.

A key aspect of the proposed approach is the combination of quick results and cumulative learning periods, so organizations shall not try to compensate for delays in the deployment of one increment by concurrently initiating the next one or by packing more features into the next increment than what can be accomplished in a 3-month period. Long or overlapping segments defeat this purpose, as they invite the same problems that plague most all-at-once implementations and work against the goal of providing cumulative episodes of learning.

8.4   Maturity models

A maturity model describes a minimum set of practices an organization must carry out in order to achieve a consistent level of performance. Central to the idea of a maturity model is the notion that to be effective, it is not sufficient to be good at one or two things, but rather, all the practices required to perform must be in place. As an example, an organization that does not plan its projects is unlikely to have a predictable performance across projects, but producing excellent plans does not accomplish much, if once the project is planned, the plan is not kept up to date or is not used to control and track work.

In practical terms, maturity models offer good advice as to which practices must be implemented together and the order in which they should be deployed. Maturity models also make good checklists to evaluate potential subcontractors. The problem with maturity models arises when they are transformed into dogma and become an end in themselves rather than a means of achieving some business goal.

Probably the best-known maturity model is the capability maturity model (CMM) devised by the Software Engineering Institute [12], but this is not the only one. Today we can find maturity models for new-product development, human resource administration, system engineering, and security. The Project Management Institute is even building one of its own [13].

The maturity model shown in Table 8.1 reflects the experience gained at Ericsson in the assessment and deployment of POs across the organization.

8.5   Communication strategy

One cannot underestimate the importance of communications in the deployment of the PO. As in any other change process, there is a need to explain the change, calm the fears of those who could be adversely affected, and get people to buy into the proposed changes.

The communication strategy must do the following:

> Identify the different audiences, their information needs, their interests, and their backgrounds in order to provide them with a relevant and understandable message;

> Decide how the information will be disseminated (i.e., meetings, presentations, Web sites, newsletters, external speakers, etc.).

People must understand why the organization needs to change. If the need for change and its purpose is not understood or intuited, people will at best temporarily comply; they will not engage in the intellectual and psychological effort required to change established routines and preconceptions. Whatever the strategy chosen, it should address the following questions[14]:

> What is wrong with the status quo?

> What is being proposed?

> How the proposed changes solve the problems associated with the current situation?

> Why employees should care?

> When employees will be affected, immediately or some time in the future?

The need for change can be conveyed by selling the pain of the status quo, or by resorting to the promise of the desired state. Different audiences respond to different arguments. In my experience what works best is a

Table 8.1   PO Maturity Model

	Expected Performance	Culture	Mental Models	Power Structure
Maturity Level	Good results are the ultimate goal.	How does the organization operate? What does it value?	What are the individual attitudes and beliefs?	Who makes decisions within the organization?
Integrated	Four out of five projects finish on-time and within budget. Productivity is in the 75% quartile.	The organization routinely uses the master plan, the resource plan, and other forecasts produced by the PO in its own planning and decision-making processes. The business value of the PO is recognized throughout the organization. Learning permeates all levels of the organization. Double-loop learning enables deeper inquiry and changes norms and assumptions. Personal mastery and team learning are encouraged.	Explicit. A system-thinking approach prevails across the organization. The "unintended" consequences of any decisions are analyzed before the decision is taken.	The PO manager enjoys a level of prestige and organizational clout similar to that of other senior managers in the organization.
Established	Three out of four projects finish on-time and within budget. Productivity is in the 50% quartile.	The portfolio steering committee meets regularly. The PO routinely exercises its authority in the prioritization and control of projects.	Explicit. The assumptions behind the decisions are brought into the open and discussed. First-order models such as "people work best under pressure" or "put more people to work to recover from the delay" are no longer automatically applied.	The regained visibility results in a transfer of power from project and function managers to senior management. The PO is empowered to make decisions within the priorities set by senior management. Decisions that could affect other projects need to be approved by the PO manager.

Table 8.1 (continued)

	Expected Performance	Culture	Mental Models	Power Structure
Defined	Two out of four projects finish on-time and within budget. Productivity is in the 50% quartile.	Common process for resource planning, project prioritization, and project reporting exists. There is a central function responsible for coordination, but often it is bypassed in the decision-making process.		
Awareness		A champion exists that voices his/her concern about the current state of affairs. Management acknowledges that something needs to be done.	Tacit. However, people start to realize that the "real world" is of their own making.	
Ad hoc	Some projects turn out right; others do not.	Hero or cowboy culture. The need to coordinate across multiple projects and establish common procedures is not yet recognized.	Tacit. When a problem arises, people justify themselves using phrases such as, "Given the situation we were in, there was nothing else we could have done." They fail to recognize their responsibility in creating the situation that comes back to haunt them.	Power resides with the project managers and the resource owners. Senior management's involvement is reactive and mostly reduced to reward, punish, or terminate the project.

Table 8.1 (continued)

	Process, Methods, Tools	Portfolio Management	Staff and Competence Development	Interfaces
Maturity Level	Describes process in place and typical tools used.	Describes what type of plans exist and how they are produced.	Describes the existence of training plans and practices for the recruitment and advancement of project management personnel.	Describes the degree of integration between the PO and the rest of the organization.
Integrated	Continuous improvement and defect prevention process in place. Models are routinely used to provide early warnings and to evaluate alternative courses of action. Integrated pipeline management tool provides accurate, up-to-date information linking resource owners, project managers, senior managers, and external partners.	Plans are evaluated using quantitative scenarios. Risks and contingencies are managed at the portfolio level. Plans are linked to their business context. Technology planning is linked to the product development plans.	Mentors are made available for guidance and support of new project managers. Knowledge sharing with other organizations is practiced. Self-actualization forms part of the project manager's responsibility.	

Table 8.1 (continued)

	Process, Methods, Tools	Portfolio Management	Staff and Competence Development	Interfaces
Established	Process for measurement, audits, and reviews exist. A project balanced scorecard is established by considering the scope, progress, cost, quality, time, and the human perspective of the project. Earned value and technical performance monitoring are fully used and understood as steering model. Integrated project, portfolio, and resource planning tools are in place.	Probabilistic analysis encourages stakeholders to take calculated risks. The status of the portfolio is known at all times.	Development paths to move from one position to another exist. Training in the critical skills required to perform the work is mandatory.	Interfaces between the PO, senior management, and resource managers are defined.
Defined	Process for estimating, project planning, resource planning, and time reporting are in place. A staged project model and standardized progress reporting provide the minimum information required to steer the project. Action items and risk issues are tracked. A central database, probably home grown, contains the organization's master and resource plans.	Master and resource plans are updated on a regular basis. Workload is kept consistent with capacity.	Job descriptions, including minimum qualifications for each position, exists. Experience in previous positions is considered, but is not the only criteria for selection. Project managers are acquainted with the nine knowledge areas identified in the PMBOK.	The interface between the PO and the projects is defined.

Table 8.1 (continued)

	Process, Methods, Tools	Portfolio Management	Staff and Competence Development	Interfaces
Awareness	Ad hoc progress reporting. Single project management tools.	List of projects with a start and end date.	Organizations at this level have difficulty retaining talented individuals.	None formalized
	Resource plans are prepared using simple spreadsheets. Multiple, usually inconsistent, versions of the same data exist.		Constant churn in the workforce diminishes its capability.	
Ad hoc			Burnouts are not uncommon.	

combination of glimpses of the golden future with flashbacks of well-known in-house episodes to which people can easily relate. Table 8.2 shows the typical content of a communications plan.

An important, and often forgotten, aspect of any communication strategy is to check that the message is getting across and properly interpreted. It is important to solicit feedback and measure the impact of PO communications to determine what is being understood and recalled, how messages are received, how receivers feel about them, and what receivers do with the information.

8.6 Limiting bureaucracy

The PO must not be seen as a heavy, bureaucratic apparatus that exists in opposition to agile methodologies and skunk work approaches—quite the contrary. The PO creates the environment for those approaches to work efficiently and consistently across many projects. The PO is there to support the projects with specialized knowledge, to guarantee that the funding and resources are in place when needed, that the project priorities remain current and their scope under control. The PO is not there to bog down the system with requests for progress reports or to be involved in the day-to-day affairs of the projects.

Table 8.2   What Information Needs To Be Provided to Whom

Audience	Message/ Information Need	Media	Frequency of Communication	Responsible	Status	Feedback
Senior management						
Line managers						
Sponsors						
Project managers						
Other employees						
Customers						

By the clever use of information technology and the empowerment of project managers, the PO should be able to function with very few personnel.

8.7   The need for the line function: How much project management is enough?

Many project management advocates herald the level of "projectization," the extent to which the organization's business is carried out through projects, as one of the attributes of an efficient organization.

This idea is largely based on the integrative and focused nature of project work, which makes it a superior form of organization when it comes to delivering concrete results in a short time. However, as T. Allen [15] notes in his book *Managing the Flow of Technology*, the structure of an R&D organization must meet two conflicting goals:

1. The coordination of the various disciplines and specialties in order to accomplish the goals of the multidisciplinary project;

2. The need to innovate and to acquire and sustain knowledge about the technologies on which the projects rely to achieve their goals.

These two goals conflict because the first one is better served by colocating all people working in a common objective and putting them under the control of a single-minded person, while the second one is fostered by keeping the project members within their functional units to facilitate the exchange of technical information and the development of new ideas.

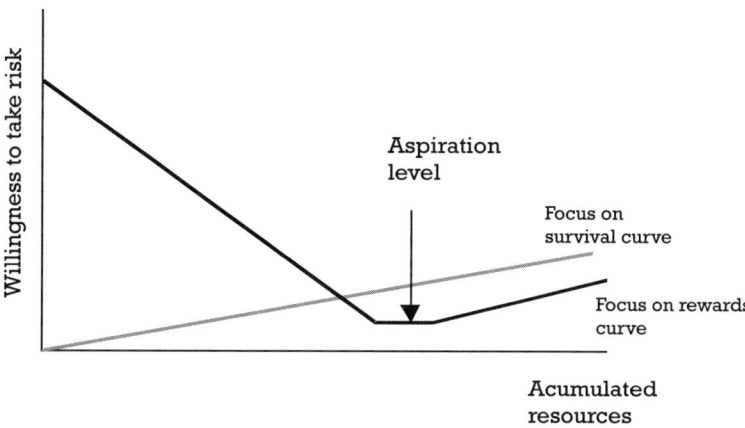

Figure 8.5 Risk behavior. (*Source:* [16].)

But this is not the only problem of working through projects. By their own nature—limited duration and resources—projects cannot afford to tinker long with alternatives, nor are project managers inclined to do so. In their study of risk-taking behavior [16], Shapira and Berndt found that managers would only take risks (see Figure 8.5), if they felt that they were not going to meet their target. Therefore, once an approach has come along that seems good enough, in all likeliness the project team will run with it and not continue to search for a better, but unknown, solution that could result in project delays or overspending. Furthermore, as the project is disbanded after conclusion, for better or for worse, the members of the team do not have to live with the consequences of whatever it is that they have developed, so sometimes they might feel tempted to take shortcuts that will come back to haunt the organization later.

For the reasons expressed above, it is necessary to supplement the transient, risk-averse, and insular nature of the projects with a line function that gives continuity to the organization and that has as a mandate not to deliver specific results by a given date, but to innovate and promote learning, tasks for which the projects are not well equipped.

8.8 Summary

Change is not easy, but it is necessary. As was stated at the beginning of this book, a more ethical and sustainable work environment is good for business, good for shareholder value, good for society, and good for us.

References

[1]   Freeman, T., *Transforming Cost Management into a Strategic Weapon*, The Consortium for Advanced Manufacturing—International (CAM-I), 1998, at http://www.cam-i.org/columns/ytransforming.pdf.

[2]   Andrews, D., and S. Stalick, *Street Smarts for Business Reengineers*, Englewood Cliffs, NJ: Prentice-Hall, 1996.

[3]   Rieley, B., *Gaming the System*, Englewood Cliffs, NJ: Prentice-Hall, Financial Times, 2001.

[4]   Beer, M., "People Express Airlines: Rise and Decline," Harvard Case Study, Cambridge, MA, 1990.

[5]   Kanter, R. M., "Power Failure in Management Circuits," *Harvard Business Review*, Vol. 57, No. 4, 1979.

[6]   Dahl, R., "The Concept of Power," *Behavioral Science*, July 1957, pp. 201–215.

[7]   Machiavelli, N., *The Prince*, 1505, translated by W. K. Marriott, http://www.ilt.columbia.edu/publications/machiavelli.html.

[8]   Northcraft, G., and M. Neale, *Organizational Behavior*, 1990.

[9]   Senge, P., *The Fifth Discipline: The Art and Practice of the Learning Organization*, New York: Doubleday, 1994.

[10] Miranda, E., L. Rosqvist, and M. Hultin, "Managing Multiple Projects," University of Linköping, Sweden, 2001.

[11] Fichman, R., and S. Moses, An Incremental Process for Software Implementation, 1998.

[12] Paulk, M., et al., Capability Maturity Model for Software, Version 1.1, Software Engineering Institute, CMU/SEI-93-TR-024 ESC-TR-93-177, 1993.

[13] Schlichter, J., *Achieving Organizational Strategies Through Projects: An Introduction to the Emerging PMI Organizational Project Management Maturity Model*, Agylon, 2002.

[14] McFeeley, B., IDEAL: A User's Guide for Software Process Improvement, Software Engineering Institute, CMU/SEI-96-HB-001, 1996.

[15] Allen, T., *Managing the Flow of Technology: Technology Transfer and the Dissemination of Technological Information within the R&D Organization*, Cambridge, MA: MIT Press, 1984.

[16] Shapira, Z., and D. Berndt, "Managing Grand-Scale Engineering Projects: A Risk-taking Perspective," Research in Organizational Behavior, 1997.

Appendix A
IDEF0 notation

The notation used to describe the PO functionality is called IDEF0. The basic element of an IDEF0 model, as illustrated in Figure A.1, is a box containing a verb phrase (e.g., "execute project") describing the activity or transformation that takes place within the box. In IDEF0 syntax, inputs are shown as arrows entering from the left side of the box, while outputs are represented by arrows exiting from the right side of the box. Controls are displayed as arrows entering the top of the box and mechanisms are displayed as arrows entering from the bottom. Inputs, controls, outputs, and mechanisms (ICOMs) are all referred to as "concepts."

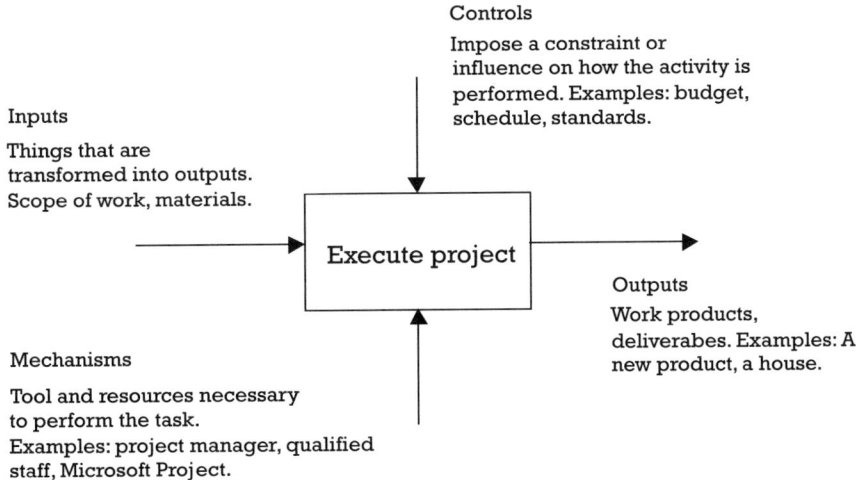

Figure A.1 IDEF0 syntax.

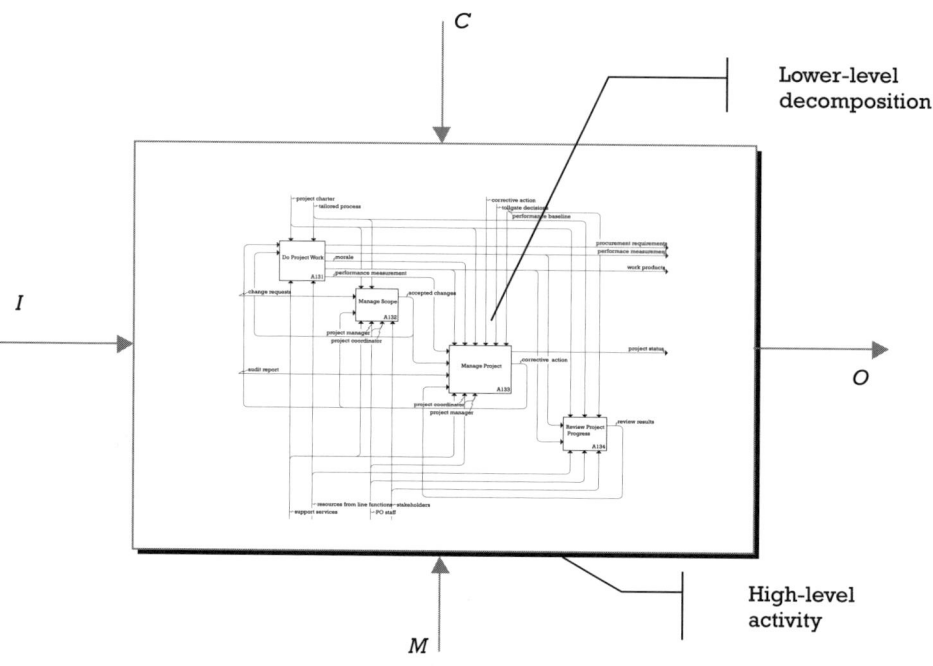

Figure A.2 IDEF0 hierarchical structure.

IDEF0 models are organized hierarchically, as shown in Figure A.2. The high-level activity as represented by the enveloping, shadow-edged box is broken down inside the box into smaller activities. The concepts entering or leaving the box at the higher level are "consumed" or "produced" by the lower-level activities. There is no need to match every lower-level concept with another at a higher level. This "tunneling" in IDEF0 terminology helps improve the readability of the diagrams by allowing for details to be shown where appropriate.

Complete information about the IDEF family of methods can be found at http://www.idef.com.

About the Author

Eduardo Miranda is currently a program director at Ericsson Research Canada. Before joining Ericsson in 1996, he was the manager of the Software Engineering Process Group for Lockheed Martin Canada. In total he has more than 20 years of experience in the development and maintenance of real-time and information management systems.

Mr. Miranda has a B.Sc. in system analysis from the University of Buenos Aires, Argentina, an M.Eng. in engineering management from the University of Ottawa, Canada, and an M.Sc. in project management from the University of Linköping, Sweden.

He has published more than 10 articles in the areas of software development methods, estimation, and project management. He is an industrial researcher affiliated with the Université du Québec in Montréal, Canada. He has also been a university professor in his native Argentina. He is a member of the Association for Computing Machinery (ACM) and the Institute of Electrical and Electronics Engineers (IEEE) Computer Society.

Index

A

Action item management, 123
Administrative support, 67
Aggregating measurements, 179–83
Aggregation structures, 179–81
    activity, 180–81
    deliverables, 180
    organization, 179
    patterns illustration, 181
    portfolio, 179
    project, 180
Analytical hierarchy process, 162–63

B

Black-Scholes equation, 167, 168
Bottlenecks, 189
Brooks, F., 38
Bubble diagram, 156
Budgeting, 90–92
    defined, 90
    goals, 90–91
    master plan and, 90
    structure, 92
    See also PO processes
Burden calculations, 68–69
Bureaucracy, limiting, 223–24

C

Capability maturity model (CMM), 27, 217–18
Capacity-versus-demand charts, 116
Change management process, 64–66
    communications, 66

configuration management (CM), 65
    defined, 64
    efforts, 64–66
    illustrated, 66
    requirements management, 64–65
    See also Support processes
Change(s)
    business-process-improvement layers of, 206
    cultural/value, 206
    lasting/effective, 205, 206
    layers of, 205–13
    managerial, 206
    mental models and, 211–13
    need for, conveying, 218–23
    operational, 206
    organizational culture and, 208–9
    people, 211
    power structure and, 209–11
    reward structure and, 206–8
Commercial success probability, 144
Communications
    path, 39
    time size effect in, 40
Communication strategy, 218–23
    message reception and, 223
    questions addressed by, 218
    tasks, 218
Complex behaved systems, 19–21
    defined, 19
    interactions, 21
    problems, 21
Configuration management, 103–4
    defined, 103

Reengineering Yourself and Your Company: From Engineer to Manager to Leader, Howard Eisner

Running the Successful Hi-Tech Project Office, Eduardo Miranda

Successful Marketing Strategy for High-Tech Firms, Second Edition, Eric Viardot

Successful Proposal Strategies for Small Businesses: Using Knowledge Management to Win Government, Private Sector, and International Contracts, Third Edition, Robert S. Frey

Systems Engineering Principles and Practice, H. Robert Westerman

Team Development for High-Tech Project Managers, James Williams

For further information on these and other Artech House titles, including previously considered out-of-print books now available through our In-Print-Forever® (IPF®) program, contact:

Artech House	Artech House
685 Canton Street	46 Gillingham Street
Norwood, MA 02062	London SW1V 1AH UK
Phone: 781-769-9750	Phone: +44 (0)20 7596-8750
Fax: 781-769-6334	Fax: +44 (0)20 7630-0166
e-mail: artech@artechhouse.com	e-mail: artech-uk@artechhouse.com

Find us on the World Wide Web at:
www.artechhouse.com